UNIVERSITY TECHNICAL COLLEGES

THE FIRST TEN YEARS

David Harbourne

Published by University of Buckingham Press,
an imprint of Legend Times Group
51 Gower Street
London WC1E 6HJ
info@unibuckinghampress.com
www.unibuckinghampress.com

© David Harbourne 2022

The right of the above author and translator to be identified as the author and translator of this work has been asserted in accordance with the Copyright, Designs and Patents Act 1988. British Library Cataloguing in Publication Data available.

ISBN (paperback): 9781915054760
ISBN (ebook): 9781915054777

Cover design: Ditte Løkkegaard
Printed and bound by CPI Group (UK) Ltd, Croydon, CR0 4YY

All rights reserved. No part of this publication may be reproduced, stored in or introduced into a retrieval system, or transmitted, in any form, or by any means electronic, mechanical, photocopying, recording or otherwise, without the prior permission of the publisher. Any person who commits any unauthorised act in relation to this publication may be liable to criminal prosecution and civil claims for damages.

Table of Contents

Author's Foreword	1
1. Lewis's story	**3**
2. Starting point 1: The JCB Academy	**8**
• Preparing to open	12
• The JCB Academy opens	16
• One year on	17
3. Ella's story	**21**
4. Starting point 2: Lord Baker and Lord Dearing	**28**
• The Labour Government after 1997	29
• Lord Baker and Lord Dearing call for 14-19 schools throughout England	34
5. Towards the 2010 general election	**39**
• Lord Dearing	43
• The Baker Dearing Educational Trust	44
6. The Coalition Government	**47**
• The Wolf Report on vocational education	50
• Becoming a movement	52
• Young Apprenticeships are abolished and the Wolf Report is accepted	55

7. Aston University Engineering Academy and Black Country University Technical College   57
- Aston University Engineering Academy   57
- Black Country University Technical College   61
- A royal visit   68

8. The next wave of UTCs   72

9. Recognizing success   78
- The Duke of York steps down   88

10. Liverpool Life Sciences UTC   90
- Edge case study   93
- Liverpool Life Sciences UTC in 2020   95
- The Studio, Liverpool   96

11. UTC Reading   98
- From start-up to maturity   101

12. Tensions and challenges   110
- Weak recruitment leads to the closure of three UTCs   114
  - Hackney UTC   114
  - Central Bedfordshire UTC   115
  - Black Country UTC   116
- Student recruitment proved challenging   117
- Exam results and student destinations   119

13. Responding to the challenges   121
- Looking for solutions   124
  - Recruitment   126
  - Running costs   130
  - Leadership and management   131
- Ministerial interest   132

14. Sheffield's two UTCs — 135
- Sheffield's second UTC — 140
- What students think about Sheffield's UTCs — 141
  - Group 1 — 141
  - Group 2 — 144
  - Former students — 146

15. Jobs for the boys? Tackling gender stereotypes in UTCs — 149
- Case study: Leeds UTC — 157

16. The movement grows — 160
- The Sainsbury Report — 164
  - Summary of main recommendations — 165
- Making ends meet — 167
- A new Secretary of State — 169
- Measuring student progress — 170
- Another closure — 172
- Research — 174

17. The JCB Academy, part 2 — 177
- Student numbers, 2018-19 — 177
- The curriculum — 178
- Ofsted — 181
- What students think about the JCB Academy — 182
  - Group 1 — 182
  - Group 2 — 186

18. Employer engagement — 193
- Case studies — 199
  - London Design and Engineering UTC — 199
  - Ron Dearing UTC — 200
- What does good employer engagement look like? — 202

19. Apprenticeships — 206
- First steps — 206

- - AUEA 206
  - The JCB Academy 207
- Broadening the scope 208
- Expanding apprenticeships at the JCB Academy 209
- Apprenticeship reforms 212
- Leila's story 213

20. Looking to the long-term 215
- The Baker clause 216
- Ministerial decisions 219
- Skewed recruitment 220
- Operational and strategic support for UTCs 222
- Marketing 225
- Accountability measures 228

21. Engineering UTC Northern Lincolnshire 229

22. Approaching steady state 236
- Reviews of government investment in UTCs 236
  - HM Treasury 236
  - National Audit Office and House of Commons Public Accounts Committee 237
- Outstanding issues 241
  - Funding 241
  - Student recruitment 243
  - Age of transfer 244
  - Multi-Academy Trusts 246
  - Teacher recruitment 246
  - T-levels and progression to higher technical qualifications 248
- Taking stock 249
  - New UTCs 249
  - Performance 250
  - Ofsted 251
- The role of the Baker Dearing Educational Trust 253
  - Simon Connell 258

23. Afterword by Lord Baker of Dorking	260
• UTCs: a century in the making?	260
• A remarkable success story	265
• Education is about more than memory	268
• Skills needs, today and tomorrow	270
• Coronavirus	274
• UTCs: the next ten years	275
24. Jodie's story	**277**
Appendix	**282**
References	**287**
Index	**303**

AUTHOR'S FOREWORD

I first met Lord Baker when I was a freelance advisor to the Edge Foundation. He and Lord Dearing asked the Foundation's trustees to help establish a new generation of schools called University Technical Colleges (UTCs).

Before long, I was advising the Baker Dearing Educational Trust as well as the Edge Foundation. I had the opportunity to visit the first UTCs very early on – the JCB Academy, Black Country UTC and Aston University Engineering Academy – and I have visited many others since then.

I recorded a long interview with Lord Baker in 2011. We talked about technical high schools set up in response to the 1944 Education Act; about City Technology Colleges, which he launched when he was Secretary of State for Education; and about his vision for a national network of UTCs.

Shortly afterwards, I read Reese Edwards' book, *The Secondary Technical School* (Edwards, 1960). Edwards was an enthusiastic supporter of post-war technical schools: as Wigan's chief education officer, he founded one of the best in the country, Thomas Linacre School, and he visited around 200 other technical schools before writing his book. It occurred to me then that someone ought to record the early history of UTCs.

Reese Edwards was in favour of technical schools. I am very much in favour of UTCs which engage and excite young people and offer multiple routes to apprenticeships, further and higher education and rewarding careers. On the other hand, Edwards was not blind to the challenges faced by technical schools and following his lead, I have tried to capture at least some of the lessons learned in the first ten years of the UTC movement.

I am immensely grateful to all the UTC students, staff and supporters who spared the time to talk to me. More than that, they inspired me with their enthusiasm, insights, ambition and sheer passion. I am indebted to Lord Baker for sharing personal papers and diary entries, as well as for all his support and encouragement – thank you! Thanks, too, go to colleagues at the Baker Dearing Educational Trust for providing access to their archives; and to the Edge Foundation and the National Foundation for Educational Research for permitting me to quote from their research and other publications.

DH
July 2020

1
LEWIS'S STORY

Lewis Clarke enrolled at the JCB Academy in Rocester, Staffordshire, when it opened in September 2010. He was 14 years old. Nine years later, he told the author about his time at the JCB Academy. Note: all interviews in this book have been edited for grammar and sense.

I moved to the JCB Academy for a few reasons. For one, I was interested in engineering and design from quite early on. I felt that at [my previous school], the opportunities for someone with that interest were limited. Secondly, I have an uncle who was working at JCB and he told me about this school that was opening specifically for engineering students, and that it was working with other businesses as well. I thought, to be honest, it would be more fun than normal school!

I was part of the first cohort so when I applied, the school hadn't been finished – there was a lot of renovation going on – but I went and met [the principal] Jim Wade and he talked us through plans for the Academy and what the aspirations were.

The curriculum wasn't exactly what I expected it to be. They told us there would be all these employer projects and that you would do your GCSEs and other essential subjects alongside that. In practice it was striking how well the subjects fitted together. I thought there would be extra-curricular projects, separate from the qualifications you came out with, but the projects were actually integrated into the curriculum. They weren't just for your own interest – they were worth something. That helped me drive to get a good outcome and to do my best on all the projects because at the end of it, I was getting a qualification, a Diploma.

A lot of learning is hands-on. You work with equipment and machinery, you do presentations and for me that was how I learnt best, whereas at [my previous school] you're in a classroom full of 30 or 40 people and there's someone at the front talking out of a textbook, which didn't suit my learning style.

There was a big difference in the hours, though. It was a big leap, going from a nine-till-three school day to eight-till-five. It was a bit of a shock at first! But on the one hand it gets you used to working hours in industry, and on the other there isn't any homework – you do it all during school hours. It was definitely good preparation for the future.

We met a lot of people from outside the Academy in key stage 4. JCB – the company – was involved in one of the projects, but I don't think that was until the end of the second year. But I emphasize to people that you don't go to the Academy simply because you want to work at JCB – it's a completely separate entity, and JCB is less involved in the Academy than you would expect. The curriculum touches on all sorts of aspects of engineering and manufacturing, not just agricultural or construction machinery. If I remember correctly, after the induction week at Harper Adams University, working with them on model 4x4s, we had projects with Network Rail, National Grid, Bosch and Rolls-Royce. We went out to those companies, did a bit of a tour, met people, then some of their staff would come on site. You got to know them and could ask questions.

We had to work in teams a lot, which I don't remember doing at my previous school. As a team, you've all got to deliver something. You start by deciding responsibilities, so one person will be a project manager who delegates tasks, sets the timing, works out who's going to do tasks together – essentially it exposes you to roles and deliverables that you wouldn't have at a normal school where you're just responsible for your own work to pass the test. I guess most students don't get that exposure to teamwork until they get to university or work, so it was good to have that early on. I found out I was quite good at taking on the project manager role, but when it came to some other aspects, it helped me understand my weaknesses. Working in a team also helped with my confidence. Doing your work on your own, and sharing it with other people – that's two very different things, isn't it?

You also have to do work experience at JCB Academy. I had three placements. One was at Zytek Automotive, looking at hybrid technology, and another was with a Ferrari restoration company. I found those myself because I thought they sounded interesting. The third one was with JCB Aviation: I got to go to a hangar at East Midlands Airport where they service aircraft for their customers, which was a really interesting type of engineering. All of this experience definitely helped shape my ideas about the future.

You go to a UTC to study engineering but because it's so broad, you need to find out which aspect of engineering you personally enjoy most and what you are good at. I initially thought I would be interested in electrical engineering and electronics, and maybe a bit of design, because I thought electricals were the future. But when it came to doing some electrical stuff at the Academy, starting with the Network Rail challenge, to be honest with you I found it actually a bit boring! I found I was more interested in design, project management and other aspects. So the curriculum and work experience helped me think through what I wanted to do in the future.

In the sixth form, I took the engineering Diploma, with an extended project qualification and A-levels in maths and further maths and AS-level physics.

The title of my engineering project was 'an advanced helicopter flight design system'. When the project started I did a work placement with East Midlands Helicopters. There had been a few helicopter crashes around that time and I noticed there wasn't anything around the blades that protected them. There was only one main rotary blade at the top, so I came up with the basic idea of swapping the one blade for two and making them adjustable. It was quite a high level project, without too much detail. That was because at the time I was interested in design and aeronautical engineering. I had to do a lot of research for that and it was all individual work. You get advice from the teacher about how to achieve a good mark, but what you do and how you present the report is entirely up to you. You have to go away and do your own research, collate it and put it together in a report, a bit like you would at university.

I liked how much independence we had in sixth form. You had to go away and do the work yourself, which was a great preparation for a degree. Compared with key stage 4, the need for self-management and motivation was up a level.

I was lucky that a lot of the staff were inspiring people. One of my mentors was Miss Boyle. She was an engineering learning mentor. She was great in one-to-one meetings, sitting you down and checking your progress. She went over and above to support me throughout the two years of key stage 4, not just with the workload but personal stuff, too. I'd also mention the maths teacher, Mr Green, and another teacher called Mr Ollis, who were both engineers by profession: they'd been in industry all their lives before making a career change in the years before retirement. Because of their background, their experience and their stories, they were quite motivational. They would try and relate topics to job roles, which really helped.

Looking back, and knowing what I know now, I think I could have put more emphasis on the engineering projects when I was at the Academy. I thought the academic side of it was the most important and perhaps I put some of the practical engineering stuff a bit to one side. But towards the end of my time at the Academy, applying for apprenticeships and so on, I really liked that I had so much to talk about. Telling the recruitment manager what I'd done I could see he was intrigued and excited because it was something new. I imagine that compared with other applicants who'd come straight from an ordinary school and done their A-levels, they wouldn't have had so much experience over and above, and that will have made me stand out. I'm sure that's what helped me get offers from Bentley, Rolls-Royce, JCB and Jaguar Land Rover – it put me in a really good position, where I could actually choose where I wanted to go.

Having said all that, I initially applied for university. In fact, I was convinced I would go straight to university all the way through my time at JCB Academy, right up until the end when my mum encouraged me to apply for apprenticeships as a backup. I found out you could end up getting your degree paid for and having a job into the bargain. I got my offer from Bentley at the end of June or early July: right up to that point I'd been expecting to go to university in September. I'd even started looking at things like accommodation, but the offer from Bentley made me change my mind.

When I compare myself with other apprentices who started at the same time at Bentley, whether it's just my personality or the extra experience I had at JCB Academy, I do feel I was a bit more comfortable getting stuck in straightaway, delegating tasks, saying 'this is what we need to do, we need to make a plan, to have a team, appoint a leader' – I just feel I was well prepared for all that. I knew what to do, where some of the others were perhaps less sure or not as confident. In essence, I think going to the Academy gave me an edge.

I finished my degree and apprenticeship in 2018 and I'm now in a full-time engineering role at Bentley. I've also started a master's degree. That's something I never thought I would do, but Bentley are willing to support me and pay for it.

My aims for the next few years are to finish my master's degree, get chartered engineer status – that's over the next three years – and then become a lead engineer responsible for whole projects and move on to the management scheme. I think that's achievable in ten years. I'd like to do a stint abroad, too.

I have spoken to lots of 14-year-olds who are thinking about going to a University Technical College. It's usually a case of, are they genuinely interested in engineering? If they are, it's a great stepping stone for finding out what engineering is like and what they are interested in. There is nothing more important than going into a career that you're interested in because you going to be doing it for most of your life: you need to be doing something that gets you up in the morning! Finding out what engineering aspect you are interested in is fundamental. The experiences you get at a UTC are not like anything in a mainstream school. I got the GCSEs and A-levels that allowed me to apply to university and sure, if that's what you want you can go to an ordinary school and get them, but you get so much more at a UTC: you visit the companies that you do eight-week projects with, and you can put all that on your CV.

I met a young guy who was thinking of going to Crewe UTC. I helped persuade him to go. I spoke to him eight months after he started and he had become this young, confident guy who had designed a seat for Bentley, packaging for a food company, this project, that project. He was what – 14, 15 years old? – and he's already done all these different projects with big, multinational companies. He wouldn't have had any of that at an ordinary school! It will make him stand out over other people, for sure.

(Clarke, 2019)

2
STARTING POINT 1: THE JCB ACADEMY

The manufacturing business, JCB, was founded by Joseph Cyril Bamford in 1945. He started by manufacturing farm trailers in a garage in Uttoxeter, Staffordshire. Anthony Bamford succeeded his father as chairman and managing director of JCB in 1975. He was knighted in 1990 and appointed to the House of Lords in 2013. Today, the company makes over 300 types of machine, including diggers, excavators and tractors, at factories in Britain and worldwide. Its headquarters and one of its manufacturing plants are just outside the Staffordshire village of Rocester, ten miles from Uttoxeter.

At around the turn of the 21st century, JCB noticed a fall in the number of young people applying for apprenticeships. David Bell, JCB's Chief Corporate Development Director and later chair of governors at the JCB Academy, had the impression that candidates applied for a JCB apprenticeship almost as a last resort: 'They'd already looked at everything else and thought, oh well, maybe I'll do an apprenticeship.' (Bell 2018)

The company had good relationships with secondary schools: for example, JCB sponsored a craft, design and technology room at Thomas Alleyne's High School in Uttoxeter. However, Sir Anthony Bamford believed more needed to be done to deepen the connection between JCB and local schools.

The head of JCB's Learning and Development Division,

Paul Pritchard, visited technical schools in Germany, France and Sweden. He was particularly impressed by Gothenburg Technical Gymnasium, an academy set up alongside Volvo's Gothenburg plant. The curriculum included both general and technical education and was devised with input from employers and universities. Students worked on 'real-world' projects based around the kind of technical challenges tackled by employees at Volvo and other engineering businesses.

Encouraged by what he had seen and with Sir Anthony's full support, Paul met representatives of local schools to discuss ideas for a teaching facility which could be shared by secondary schools across Staffordshire and Derbyshire. Students would go there for a day or half day a week to learn engineering theory and skills. However, the idea did not take root. Objections included the disruption to normal school timetables and concerns about health and safety.

There was a second visit to Gothenburg. This time, Paul was joined by David Bell; Peter Mitchell, head teacher of Thomas Alleyne's High School; and Keith Norris, principal of Burton College. They, too, were impressed by the Gymnasium and its students, who appreciated that the curriculum prepared them both for work and for higher education.

Sir Anthony Bamford was very enthusiastic about the idea of establishing a similar academy in Rocester, and the time was right: Andrew Adonis, Parliamentary Under-Secretary of State at the Department for Education and Skills (DfES), was actively seeking sponsors to open new schools under the Labour government's academies programme. David Bell recalled:

> We went to the north-east to meet Peter Vardy and visit one of his academies. Paul Pritchard and I were absolutely bowled over with what we saw and the atmosphere in the school. That got us thinking about it, and it came together with the idea of raising aspirations and boosting the profile of engineering and manufacturing...
>
> We were responding in part to the Leitch report on skills [Leitch, 2006], and the 14-19 agenda was on the table [see chapter 4, below]. At the very beginning, before we decided what to do, we went to see some 'standard' academies, which were 11-18; but once we

got to the point where we were trying to stimulate interest in engineering and respond to the Leitch report, we never seriously looked at 11-18.

We met Tony Blair and Andrew Adonis in Birmingham and they said yes, it's a brilliant idea – love it – and sent us to see an official in the academies division [of the DfES], who was fairly positive about it.
(Bell, 2018)

Lord Adonis mentioned JCB's plans in the House of Lords in June 2006:

Only three days ago, I was talking to people at JCB [who] wish to develop … an academy specifically to teach construction and engineering skills to 14 to 19 year-olds … JCB is attracted to the [academy] model because it provides certainty in developing the model over time.
(HL Deb 21 June 2006, col.863)

JCB identified suitable premises for an academy, a cotton mill built in Rocester in 1781-2 by one of the Industrial Revolution's leading figures, Richard Arkwright. Lying unused since its closure in 1985, Tutbury Mill would need a lot of investment, but would be a powerful symbol of continuity and innovation in engineering and manufacturing. Once the project was approved, the old premises were restored and linked to new, purpose-built classrooms, workshops and ancillary accommodation.

JCB did not work alone. Other early supporters included Rolls-Royce, Toyota, Bombardier, Bentley Motors and Network Rail. Harper Adams University and Thomas Alleyne's High School were involved, too. All agreed that the new academy should recruit students at 14 and 16 – not 11 – from a wide catchment area covering Staffordshire, Stoke, Derby and Derbyshire, and offer a broad curriculum alongside engineering. On leaving the academy, students would progress to further education, apprenticeships or university.

In March 2006, the government included the JCB Academy in a list of 100 academies that had already opened or were in the pipeline:

ACADEMY PROGRAMME REACHES HALFWAY MARK
Today the Prime Minister announced there are currently 100 Academies open or in the pipeline – half-way to the established target of 200 Academies...

Education Secretary Ruth Kelly said: "Academies have come a long way in a very short time, considering the first three opened in 2002, and it is good news for some of our most disadvantaged communities that sponsors and local authorities across the country have been inspired to embrace the exciting opportunities they bring.
(Department for Education and Skills, 2006)

By then, the JCB team led by Paul Pritchard and David Bell was ready to submit an application to the Department for Education and Skills:

When we started looking at the expression of interest questionnaire, there was nothing that fitted what we wanted to do. It was full of questions about the 'predecessor school' and area of deprivation. We could fill about a tenth of the form in because it was mostly irrelevant to what we wanted to do.
(Bell, 2018)

The first expression of interest was rejected on two grounds: the village of Rocester was not classed as a deprived community, and JCB planned to recruit pupils at 14 and 16: academies were expected to cater for the full 11-19 age range.

A number of factors led officials to change their position and accept JCB's proposals in 2007. First, recruitment would extend over an 18-mile radius, encompassing many deprived wards. Second, a three-tier system of primary, middle and high schools operated in some parts of the catchment area: unlike other parts of the country, transfer at 11 was not the norm in those areas. Third, engineering and manufacturing employers were enthusiastically involved in developing the new engineering Diploma (described in chapter 4, below), which would provide the backbone of the JCB Academy curriculum. Fourth, the Government had signed up to the Leitch

Report, *Prosperity for all in the global economy – world class skills* (HM Treasury, 2006), which said that without fresh investment in skills, the UK economy would lose ground to its competitors. The proposed academy would contribute directly to meeting the aims of the Leitch Report. In addition, the Department now appreciated that questions about the JCB Academy's non-existent 'predecessor school' did not apply.

The JCB Academy Trust was formally registered with the Charity Commission in August 2007. Its aims were:

> To advance for the public benefit education in the United Kingdom, in particular... by establishing, maintaining, carrying on, managing and developing a school offering a broad curriculum with a strong emphasis on, but in no way limited to engineering, manufacturing and international business.
> (The JCB Academy Trust, 2007)

PREPARING TO OPEN

Jim Wade was appointed principal-designate of the JCB Academy in October 2008 and took up his appointment in January 2009. He applied for the position because of a long-standing interest in vocational education:

> My background is business education. I taught at one of the first schools to deliver BTEC [Business and Technology Education Council] qualifications. Later, when I was head of sixth form, we introduced GNVQs [General National Vocational Qualifications], so vocational programmes had always been at the forefront of my experience. I saw the JCB position advertised and thought, that looks interesting!
> (Wade, 2018)

As soon as his appointment was confirmed, Jim Wade met Paul Pritchard to discuss plans for the new academy. At that point, the governors planned to select students at 14 on the basis of their aptitude. Jim warned that the government was unlikely to agree to any form of selection. Sure enough, DfES confirmed that recruitment had to be comprehensive; if the academy was oversubscribed, places would be allocated by a form of random allocation such as a lottery and not by aptitude.

The curriculum was to be based on Diplomas, with a strong emphasis on employability skills and the direct involvement of employers in delivering cross-curricular projects. It was expected that a majority of students would remain at the academy after the age of 16, while a minority would leave to take up apprenticeship places or courses at other schools and colleges. At that stage, the academy would not itself offer apprenticeships, though there were plans to do so in later years. Speaking about the curriculum, Jim Wade said:

> The governors knew what they wanted, but were not necessarily clear about how to get it within an educational context – it wasn't their area of expertise. Working with Harper Adams University, we devised an off-road challenge, which involves designing and building a remote-controlled 4x4 car. We mapped the project against Diploma specifications and showed how it could be delivered. I employed maths and English specialists to map the project against those subject specifications as well. I presented that to the academy board to show how we would deliver the specifications and involve our partners in the curriculum.
> (Wade, 2018)

The JCB Academy described the 4x4 challenge in these terms:

> The 4x4 challenge is the first major activity for the new Y10 students and commences with a five day residential at Harper Adams University. During their stay, students engage in a wide range of workshops with staff from Harper Adams, including chassis design, suspension, power and transmission, all of which develop their knowledge and understanding of 4x4 vehicle design. As well as the technical workshops, the students participate in various activities on site, e.g.

orienteering, dodgeball, rounders and team building as well as off-site activities such as bowling. This is a fantastic opportunity for the students to get to know each other, as well as some of the Academy staff.

After returning from Harper Adams students work in teams on the challenge which is to design and manufacture a 4x4 vehicle. Students design their chassis in 2D and then use the laser printers to manufacture the parts. They then build the cars for testing on the track, making modifications where required in order to improve performance, ready for judging by Harper Adams staff.
(The JCB Academy, 2013)

Wade explained the 4x4 challenge to employers and invited them to come up with their own curriculum projects. He said:

We ran a series of three-day conferences, each of them attended by two or three people from three different employers. With each of the three employer groups, we had someone representing the academy and someone from OCR [Oxford, Cambridge and RSA Examinations], and we employed maths, English and science consultants to help as well. Basically, we locked each group in a room to develop their project ideas. From time to time, they came out of their rooms and shared their ideas and where they had got to. They sparked off each other's ideas – it was a fantastic process.

What we're able to do is transfer ownership of the projects to the employer. Still to this day, each employer owns their own project – the Rolls-Royce project, the Toyota project, and so on. People within those organizations feel responsible for their challenge.

If I reflect on my experience in other schools, we had a tendency to invite employers in to do something – come and speak to the students or do something very specific with them. But here, they feel they *own* the curriculum. By going through that development process, they took ownership and wanted their project to be great – better than everyone else's!
(Wade, 2018)

The academy chose to work with the qualification awarding organization OCR, which offered substantial support throughout the development phase and beyond. Jim Wade said:

> Not only did OCR devote considerable time to helping us map the curriculum, but there were occasions when they agreed to change the Diploma specification as a result of some of the work we did together. It helped us, of course, but it also benefited them. I don't think they'd ever worked with the range of engineers that we worked with here: on occasions, changes were made to the specifications because engineers pointed out they were actually wrong.
> (Wade, 2018)

The governors were happy with progress, but wanted assurances that the emerging curriculum would meet all key requirements. Professor Matthew Harrison, then Director of Education at the Royal Academy of Engineering, carried out a review and confirmed that the process was on track to deliver a sound and effective curriculum.

Jim Wade also devised a staffing model, greatly helped by generous government funding for institutions delivering the Diploma. His model divided students into house groups. Each house group of 40 students was supported by four staff: a qualified teacher, a support assistant and two learning mentors drawn from an engineering background. Recruitment was less challenging than it might have been, partly because of the recession affecting the economy at that time: there were relatively few job opportunities for people seeking work in engineering and manufacturing.

Jim Wade recalled that recruiting the first intake of students was a greater challenge:

> In the first year, we weren't allowed to go into any of the other schools and they wouldn't distribute our literature. Instead, we did a series of roadshows in each of the towns around the region. We had a van and a set of kit – plasma screens, portable engineering equipment and so on – and went round church halls, sports centres and the like. We leafleted the areas where roadshows were held. In some places, we had a

> great turnout; in others, you could see the tumbleweed rolling through the venue as three people stood at the front waiting for the presentation to start! JCB [the company] got some of their apprentices to come along, and governors came too to answer questions.
>
> (Wade, 2018)

THE JCB ACADEMY OPENS

In the event, the academy exceeded its initial recruitment target for year 10 and allocated places by lottery. The JCB Academy opened in September 2010 with 120 students in year 10 (age 14) and 50 in year 12 (age 16), drawn from 38 different secondary schools.

Some year 10 entrants chose to join the JCB Academy as much because they wanted to leave their old school as because they were attracted by the engineering curriculum: reasons included bullying, not fitting in and poor behaviour. To these students, the academy represented a fresh start.

Academy staff were surprised to find that an unusually high proportion of students – perhaps as many as four in ten – had special educational needs, particularly relating to dyslexia. Less surprising was the difficulty of recruiting girls to an academy specialising in engineering: 11 girls joined as part of the year 10 intake, and seven joined year 12.

The teaching day started at 8.30 am and ended at 4.00 pm. All school work took place during the day: no homework was set in years 10 and 11. At 4.00, students could choose from a range of sports and other activities – the Duke of Edinburgh Award, for example – before leaving for home at 5.00 pm. The school year was longer than usual, too.

Projects delivered in the 2010-11 academic year were supported by JCB (vehicle maintenance), Toyota (manufacturing engineering), Network Rail (engineering applications of computers), Rolls-Royce (engineering design) and Bombardier (innovation, enterprise and technological advance).

In year 10, students studied for the engineering Diploma, plus GCSEs in English, maths, science, German and ICT. They also covered other national curriculum requirements including religious education, citizenship, enterprise and careers education. Post-16 students studied for an advanced Diploma in either engineering or business. A-levels were also offered – some of them in partnership with Thomas Alleyne's High School in Uttoxeter.

The Academy was officially opened by Their Royal Highnesses The Prince of Wales and The Duchess of Cornwall in February 2011. Prince Charles said:

> Meeting the students was particularly interesting. Some of them have to get up at half past six in the morning in order to be here on time, and they go on working until five o'clock in the evening. They are working incredibly hard, in a remarkable atmosphere, with remarkable teachers, tutors and mentors … I think this Academy – and I hope, others like it – will make an enormous difference to this country's future.
> (The Royal Family, 2011)

ONE YEAR ON

The Baker Dearing Educational Trust (BDT) published a report to celebrate the first year of the JCB Academy (Ware, 2011). The author, Jane Ware, set out the academy's vision and purpose:

> The vision of the sponsor JCB is to develop employable young people with positive attitudes, emotional intelligence, intellectual horsepower and appropriate competencies; to achieve excellence in academic and vocational education; and then provide a stimulus to improve provision across the region for engineering, manufacturing and business skills …
>
> The defined purpose is of course to find out about engineering. A secondary purpose, however, has emerged. Certain students who have misused their time at a previous school have taken the opportunity of the non-standard age of transfer at 14 for a second chance. At the academy they are treated in a more adult way. There is a singleness of purpose and a greater amount of the learning is done in a practical way. This has suited those 'second chancers' and the students agree with

parents and staff that it has transformed their outlook and prospects.
(Ware, 2011, pp. 5-6)

One of the projects described in the report was devised by Rolls-Royce and concerned the design and manufacture of a piston pump for a jet engine:

> Rolls-Royce graduate apprentices designed and manufactured a pump rig which has a series of interchangeable parts. This allows the students to change the bore and stroke of the piston pump and to prove the best design.
> The pumps are then modelled using Siemens NX7.5 software. This allows full 3d modelling and animation. From this, students can check parts and produce drawings. Following a visit to the factory to see how the real ones are made, students are helped by Rolls-Royce engineering apprentices in the workshop to realize their pumps. All pumps are tested and must be within tolerance.
> The unit is supported by senior Rolls-Royce staff who present not only how a jet engine works but also responsible business practice, planning and aspects of local and international business.
> (Ware, 2011, p.26)

The report noted the diverse backgrounds of the academy staff (Ware, 2011, p.10). One had previously worked in an engineering and mentoring role in the Royal Navy. An engineering team leader had a degree in industrial design. Another was an engineer with Proctor and Gamble. Others came from more conventional teaching backgrounds. The wide span of prior experience enabled cross-fertilization of ideas and approaches to teaching.

A surprisingly high number of students – 36 per cent – had special educational needs (Ware, 2011, p.20). There were seven or eight dyslexic pupils in each class. Seven students joined the academy because of behavioural problems at their previous schools.

Students were offered mentoring. Engineering Learning Mentors (ELMs) attended all engineering lessons and workshop sessions, as well as English, maths and science lessons. They sat with students at lunch, too. ELMs reported to parents by email every second week. Ware wrote:

> After 23 years as an aircraft engineering mechanic working on radars and radio systems, Mark Spooner is an ELM who brings to the academy an understanding of engineering as a career and experience of mentoring.
>
> Another staff member from an engineering background is Engineering Team Leader, Paula Gwinnet. Although on course for 'A' grades in mathematics and science, she felt compelled to pursue engineering and left school at 16 to joining a small machine shop. She then went to technical college and on to Sheffield University to study mechanical engineering. Determined to make a difference to engineering education in the UK, she joined the academy to lead a team of mentors and a learning support assistant to deliver the Diploma curriculum to a House year group, 40 Year 10 students.
>
> (Ware, 2011, p.21)

Jane Ware interviewed Ella Pilsworth-Straw about her first year at the JCB Academy (Ware, 2011, p.15). Ella joined the academy at the age of 14.

Jane: Are you glad you came here?

Ella: It's been good and I enjoy it a lot. In most challenges we have had team working and it involves more competition than I thought it would, especially between Houses. I do like going in the workshops for practical work but I am glad it is not every single day.

Jane: What was your favourite challenge?

Ella: The 4x4 challenge. I had a bit of knowledge but it was helpful that they started from scratch at Harper Adams. We learnt about surface area, what shape we were going to have it and steering.

At first our team thought we would make the vehicle quite simple. But we did a presentation and the lecturer suggested that if we made the chassis as small as possible, it would make it better so we decided to have the battery pack on an angle to make the chassis smaller. This meant it would have a smaller turning circle and it would be easier to manoeuvre. So it was quite different to most of the other groups' vehicles.

Jane: And do you think it produced a better product?

Ella: I think it might have done if our car worked completely properly but our steering mechanism broke. We did a lot to make it stronger but it just made it worse in the end.

Jane: How do you spend the extension time?

Ella: Netball, art and drawing – which is good practice because you have to do a lot of sketching, particularly for the Rolls-Royce challenge. I do triple award science on a Thursday night.

Jane: Plans for the future?

Ella: I'm interested in biomedical engineering and fake limbs. I've just been on a taster course at Manchester University in material science where we looked at x-ray machines, learnt about selecting materials and did group projects.

(Ware, 2011, p.15)

3
ELLA'S STORY

Eight years after her interview with Jane Ware, Ella Pilsworth-Straw spoke to the author about her memories of the JCB Academy and what she had done since leaving.

I went to a school for girls before transferring to the JCB Academy. In some ways it was very traditional: they expected you to choose more conventional subjects like cooking, art and law rather than maths and science, which had always been my favourite subjects. I knew even then that I wanted to go into engineering, which is why I made the leap.

My step dad's an engineer, but if I'm being completely honest I never really knew exactly what he did! It was more the fact that I liked science, and especially maths – I liked the application of it as well as the pure side of the subject. It stemmed more from that, really, rather than any kind of influence from my family.

I was part of the very first intake at the JCB Academy. It was a huge change for me, going from an all-girls school to what was almost an all-boys school. It was actually fine, even though there were only a few girls. It felt a little challenging at first, but the girls stood their ground and we fitted in well. We were all there for the same reason – we had similar interests and it was quite easy to get to know everybody.

In the summer before we started, we had an induction day where we chose who we wanted to be in a house with. I turned up knowing absolutely no one and it was quite daunting. We were put on tables, and I got to know some of the other girls, which was nice. Going to Harper Adams University for a weekend at the start of term was super-helpful, too. We were in our houses, but got to know people in the other houses as well. Harper Adams was so much fun that when I was in year 13 I actually went back as a helper for the new year 10s!

The Harper Adams 4x4 project introduced us to the style of learning where we worked on projects over eight or 12 weeks – that was the basis of our Diploma. It settled us in and got us used to learning at JCB. Because it was fun it made it easier for us all to get along and make friends.

I really appreciated the time at Harper Adams. It set high expectations. And the curriculum lived up to those expectations. It was great that we had such a variety of projects in the engineering Diploma as well as a variety of GCSEs. You didn't get the same range of options for GCSEs as you would in a normal school, but you knew that before going into it. The academy still took account of whether you were more academic or more hands-on: you could do more engineering tasks in the workshop, for example. As the first ones in the academy we could be seen as the guinea pigs, but I think we all learned together – students and teachers alike – and it made us all quite close as a unit, as a year, which helped us get the most out of our time in key stage 4.

Every time we did a challenge with a company, we worked in different teams which meant you got to know different people. That really helped me understand that you are not always going to work with your friends, which is really important. Some people think 'I'm just going to be with my friends' – especially at that age – and that it's all going to be great fun, but we learned how to adapt to different people's needs and the contributions they make to the group, and how to enable everyone to work well in a team. We learned that not everyone has the same outlook as you, not everyone is going to work as hard as you or in the same way as you. Learning how to cope with that was super-important because, going into the world of work, you are working with all types of different people and it's important to be approachable and adaptable. It was extremely useful at university as well.

Before I chose my A-level options, I was already looking at the kind of courses I wanted to take at university and basing my choices on that. I was lucky that I knew what I wanted to do from quite early on. I know not everybody was the same, and I feel lucky that I knew exactly what type of engineering I wanted to go into – bioengineering. I knew I didn't want to do standard mechanical engineering. I was never at my best in the workshops and I think that's when I started to do my research. I also love the idea of helping people: I think that's where it stemmed from initially.

Weirdly enough, it was a sector of engineering we didn't fully cover at the JCB Academy, but I knew for definite that I wanted to do it, so I did some extra courses outside school time – materials engineering, for example, which is a big part of bioengineering – and my appreciation for it just grew and grew. We had to do a project at the end of year 11. You could do it on anything you wanted and I chose bioengineering. I also did a course at Cambridge University – a Nuffield project – over one summer holiday. One of my teachers set me up for that. Teachers were really accommodating like that.

Engineering is actually about solving problems; taking real-world issues and creating solutions. I think that aspect really shone at JCB Academy. It wasn't just about hands-on experience in the workshop. Obviously, that's helpful because even if you are doing a project which leads to a written report, it's still good to break down the issue and even literally break something down to work out why it's not working.

Engineering is much broader than many people think – it's about the application of principles, of physics, of maths, to solve problems. Too many people think engineering is the same as being a car mechanic, but that's a misunderstanding. We need to raise awareness of what it's really about and help people choose engineering for the right reasons rather than what they presume it might be. It's not just about working on your own or working with machines, it involves working with other people. Everything I've done has involved working with a team.

We had to do work experience every year as part of our Diploma. It felt quite daunting because when I started I was – what, 14, 15? – and going out to do work experience in a big engineering firm, especially as a girl, where almost everyone else was a man – it was a big deal. But it was very eye-opening. Everyone was very encouraging. The careers team at the JCB Academy helped a lot but also encouraged you to be independent. It's all very well mollycoddling you, but that doesn't really help in the long run. They showed us how we could go about getting our own work experience, which was very helpful. It made us more confident and independent, which was very beneficial as we got older. It's helping me even now to apply for jobs, knowing how to present myself.

Obviously a lot of the work experience was not in bioengineering, but even that was helpful in showing me careers that I didn't want to do. That was just as helpful because I needed to narrow down what I wanted to do. And when you were on work experience, someone always came out to see you and make sure you were okay. There were lots of opportunities to discuss work experience when you got back to the academy and to think about what you'd got from the experience. For some people it was really inspiring and made a big difference to the choices they made. It was very purposeful – a big part of the academy's approach to the curriculum. It wasn't just an add-on, as it is in many schools.

Doing so much work experience also gave me a lot to put on my CV and to talk about in interviews. It can be a vicious cycle: you've got the grades, you want the job, but you don't have the experience. You need someone to give you that break. It was really helpful to have so much work experience – it helped break the vicious cycle.

Even more broadly, the careers team are very encouraging and hands-on, sitting down to discuss what you want to achieve and what you want to do. You are never afraid to ask for help. If I had told my previous school that I had wanted to do an apprenticeship, it wouldn't have been exactly frowned upon, but I would definitely have been encouraged to go to university instead, whereas at JCB, every option, every career idea, met with the same response – 'right, let's get you there'. That was something very special about the JCB Academy, which I don't think every school has.

It's hard to pinpoint individual members of staff as especially helpful, because they were all amazing. I know it sounds a bit cheesy, that I would go back in a heartbeat if all the same teachers were there: I would do it all again. Mr Wade, the principal, was extremely helpful – and still is! If I ever need anything, he's always there. When I was looking for an industry placement, he said, 'right, I'll see what I can do, I'll see if I've got any contacts'. Another really important teacher was Miss Gwinnett, our head of engineering: it was really inspiring to know that a woman can have a great career in engineering.

And then there's my maths teacher, Mr Clarke; Mr Greene; Mr Ollis ... Mr Ollis was one of the most incredible teachers I've ever had: he chose to be a teacher because he wanted to give back. Like a lot of teachers at the JCB Academy, he hadn't been teaching all his life: they had other careers and experiences. Mr Ollis went above and beyond not just for me but for all of us. In the sixth form, a bunch of us did extra A-levels. Actually, you can't do that anymore because it was a crazy workload, but Mr Ollis went out of his way to help us when he could. In fact it was Mr Ollis who got me the placement at Cambridge University and I will be forever grateful for that.

In years 10 and 11, we had a mentor. They sat in on lessons in your house – that's a group of about 40 – but you also had a teaching assistant on a ratio of about ten students to one assistant. You had weekly catch-ups and you could talk to them about anything at all. My mentor was called Miss Norcup. Even in years 12 and 13 she made time for me whenever I was stressed about something – university applications, or whatever. She was fully invested in how I was going to do in school and in life. Even now, when I go back to the academy, she's the first person I look for: I give her a hug and she asks how I'm getting on. Thinking about my previous school, when you went to parents' evenings, teachers had things written down that they wanted to tell my parents – they knew what I was doing in lessons – but I definitely felt more valued at the JCB Academy. So to go back to my earlier point, it's really hard to single people out when I really did love all of my teachers!

When I was 16, I got offered an apprenticeship at Toyota. We did a challenge with them and my team won. The head of manufacturing watched us and he headhunted me. I politely turned it down because at 16, I just wasn't ready. I'd always wanted to do A-levels because I am a very academic person, and I knew where I wanted to go. Toyota would have been amazing, but at the same time I owed it to myself to do what I was really passionate about.

So I did consider other options before going to university, but to be honest there aren't many apprenticeships in bioengineering. In fact, there aren't many uni courses. I was one of only a few of my friends who chose to go to university – the others chose apprenticeships and most of them are thriving. I've just finished university after five years, including a year in industry and completing a master's.

I would say, 100%, that I was better prepared than other students on my course. I had done a lot of independent work already and I'd also worked a lot in teams on engineering problems. A lot of the other students had come from ordinary schools where they'd just done – you know – maths, physics and biology, and they didn't have the same report-writing skills and experience I had. I knew some of the applications already, too – it wasn't just about pure maths, I knew how to apply mathematical principles to the engineering side of it. I think it might have been a bit of a shock for some of the others who were not used to working in teams or doing a lot of independent work. At the JCB Academy, it's up to you to hand in your reports and meet deadlines: a lot of people from other schools were not used to that because they were handed things on a plate. When they arrived at university, they were not used to managing their own time: someone always managed it for them.

While at university, I did work experience during the summer because I knew it would be extremely helpful. Not everyone at uni thinks about their future in the same way: many people are thinking about 'now'. But everyone is going to have that degree when they finished; everyone is going to have the same result, or very similar. A 2:1 or above is what you need to get a good job. But if everyone goes in with the same starting point, it becomes a question of what you have that's extra. So I did a placement at Nottingham University one summer, which I really enjoyed. Actually it also made me realize I didn't want to do a PhD, but go straight into industry.

Then of course I had a year in industry as well, which only a few of us did. Only about seven people out of all of us – there were 50 of us in the year – did the year in industry. Admittedly, they were hard to get and I was extremely fortunate to go where I did – DePuy Synthes, owned by Johnson and Johnson – but I thought it was really important to have a year in industry. Other people thought it was crazy, taking such a long time to complete my degree, but I thought it was better to get experience now, when I can, because employers want you to have experience. Otherwise, it's that vicious cycle: you need experience to get a job, but without a job you can't get experience. I thought I would either love it there, or hate it so much I wouldn't want to do orthopaedics, but I ended up absolutely loving it! I ended up staying even longer than a year and I'm actually going back there in September to join their graduate scheme. Without that experience, I don't think I would have been able to get past the interview because I wouldn't have been able to tell them about experiences I had had in bioengineering applications.

Looking ahead, I would very much like to progress and become a valuable member of the team. I know I was already a valuable member of the team during my placement year – don't get me wrong, I didn't just make cups of tea! I was fully involved in everything and made presentations to interdisciplinary teams. I felt I was making a difference and helping people learn something new.

It would be amazing if, in ten years' time, I was seen as the go-to person in a specialist area, to a high level. I think that would be really impressive. But I'd also like to look into the managerial side: I really like the idea of a leadership role and progressing myself. DePuy are very committed to developing people and I really think I will be able to make the most of my abilities there, though I know that as a female in a mainly male workplace I'll have to work hard to make my mark. And of course I'm still young – though in fact, there's a lot of younger people there now.

I want to get across my appreciation for the JCB Academy. Whenever a few of us go in, especially from that first year group, we always want to make it clear how appreciative we are for our time there. The teachers who are still there from our time can't wait to talk to us – they are really interested in what we're doing now. I would do anything to advertise and promote the JCB Academy.

I know some people have tried to bash UTCs. I want people to know that going to the JCB Academy has really shaped my life. I know I had advantages before that, but it really was the academy that made me who I am. They call you a student, not a pupil, to show their respect. You work normal business hours, not a short school day, because they want you to know what it's like when you're working. You wear a suit, not a normal school uniform, because they want you to feel important. And we studied for the Diploma, which was amazing. Doing applied subjects doesn't mean you're less intelligent: it's just a different way of learning. And we got so many transferable skills from studying engineering – problem-solving, working in teams, communicating – which you can adapt and apply throughout life.

(Pilsworth-Straw, 2019)

4

STARTING POINT 2: LORD BAKER AND LORD DEARING

Kenneth Baker was Secretary of State for Education between 1986 and 1989. His reforms included launching the National Curriculum and introducing four 'key stages': school years 1 and 2 (key stage [KS] 1), years 3-6 (KS2), years 7-9 (KS3) and years 10-11 (KS4). He also launched City Technology Colleges (CTCs), new secondary schools which – unlike other state schools at that time – were directly answerable to the Secretary of State for Education rather than local authorities. He said:

> I was determined to build on the idea of technology learning and within a few months of becoming Secretary of State, I established the concept of City Technology Colleges. They would teach using technology, be independent of local authorities, have control over their budgets and involve businesses.
> (Baker K., 2011c)

The original plan was to open 20 CTCs, each offering a curriculum matched more explicitly to the needs of employers, who would be directly involved in running them. An independent City Technology Colleges Trust was set up to find potential sponsors and supporters, and to help CTCs establish themselves. Sponsorship was secured mainly through the enthusiasm of individual entrepreneurs including

Harry Djanogly, Stanley Kalms, Peter Vardy and Philip Harris, and philanthropic organizations such as the Mercers' Company.

Kenneth Baker's successors as Secretary of State for Education did not share his enthusiasm for CTCs, some of which got off to a rocky start. In the event, only fifteen were opened.

However, that was not the end of the story. CTCs got over their teething problems and started to make a difference. Looking back on the programme, Lord Baker said:

> I wanted 20 CTCs but was moved to a new job before the programme was completed. In the end, we got 15 off the ground. They are still some of the most successful schools in the country.
>
> What CTCs showed me was that taking schools out of local authorities would release a huge amount of energy, imagination and ideas. If you gave them a chance at the local level teachers would show a great deal of creativity.
> (Baker K., 2011c)

Kenneth Baker left the government in 1992 and stood down from the House of Commons in 1997. Shortly after the 1997 general election, he was granted a peerage and joined the House of Lords as Lord Baker of Dorking.

THE LABOUR GOVERNMENT

Tony Blair became Prime Minister in 1997. The following year, he appointed Andrew Adonis as his education advisor.

Adonis was convinced that secondary education needed further reform. He said:

> One statistic shocked me above all. In 1997, fewer than half of all state comprehensives – *fewer than half* – achieved a decent school leaving standard for more than one in three of their sixteen-year-olds, defined by the basic yardstick of five or more GCSE passes at grades A* to C including English and maths.
> (Adonis, 2012, loc. 672)

At first, Adonis lacked a clear, worked-up plan. Then in October 1999, he visited Thomas Telford School, a CTC co-sponsored by the Mercers' Company and Tarmac, where he found …

> … an ethos of achievement and excellence akin to the best grammar schools, yet for pupils of all abilities and backgrounds with a modern curriculum. There was a huge sixth form sending dozens each year to university. The teachers were highly qualified, ambitious and passionate. Every pupil appeared to be succeeding at something.
> (Adonis, 2012, loc. 980)

In March 2000, the Secretary of State, David Blunkett, made the first public commitment to a small-scale, experimental academies programme based on many of the principles underpinning CTCs. Although the announcement was low key, it was to have a profound effect on education in England.

David Blunkett's successor as Secretary of State, Estelle Morris, published a white paper, *Schools: Achieving Success* (Department for Education and Skills, 2001). It included a commitment to open 20 academies by 2005. It also made the case for greater curriculum flexibility in key stage 4, the aim being to match the curriculum more closely to the interests and aspirations of individual pupils. The new curriculum would be supported by a new, over-arching award to recognise achievements in both academic and vocational studies between the ages of 14 and 19.

In 2003, the government set up a 'Working Group for 14–19 Reform', chaired by a former Chief Inspector of Schools, Mike Tomlinson. Announcing the working group, the government drew attention to persistent pupil under-achievement, absence and poor behaviour:

> Nearly half of young people still do not achieve five good GCSEs at school. More still do not reach that standard in English and mathematics. And one in twenty leaves without a single GCSE pass. These data reflect a deeper problem. Too many young people truant in their last two years of compulsory education

[years 10 and 11]. And the behaviour of some who turn up makes it hard for teachers to teach and others to learn …

For too long, we have thought in terms of two phases: 11–16 and 16–19 … To think in terms of two phases is no longer helpful or meaningful. This is why we wish to develop a 14–19 phase.
(Department for Education and Skills, 2003)

The Tomlinson Report – or more precisely, *14-19 Curriculum and Qualifications Reform: Final Report of the Working Group on 14-19 Reform* – was published in 2004 (Department for Education and Skills, 2004b). It contained a range of recommendations, but one was especially important: 'The existing system of qualifications taken by 14-19 year olds should be replaced by a framework of diplomas at entry, foundation, intermediate and advanced levels' (Department for Education and Skills, 2004b).

The Secretary of State supported the idea, but the Prime Minister did not. The BBC's education correspondent noted that on the very day the Tomlinson Report was published, Tony Blair announced that A-levels and GCSEs would stay (Baker M., 2005). Sir Mike Tomlinson said later:

> When the interim report came out, everybody was onside – absolutely everybody, across the education sector and employers alike. I got the approval of the then Secretary of State and the junior minister directly responsible, and cross-party support. The final report then emerged, not vastly different in outline but with somewhat more detail and somewhat greater understanding of certain matters.
>
> I thought the [Prime Minister's] proper, judged response would be to say that until he had read the report in full and considered it with colleagues, he would be making no judgement about the way forward. Instead, he voiced his opposition to our main recommendations, taking the Secretary of State rather by surprise. It was a great shame, because we had everybody lined up.
> (Tomlinson, 2017)

Having rejected the central recommendation of the Tomlinson Report, the government developed new 14-19 Diplomas to sit *alongside* existing GCSEs, A-levels and other qualifications. Each of the 14 Diploma 'lines of learning' would be linked to a sector of the economy such as engineering, creative and media or construction and the built environment, and be offered at three levels.

Every Diploma included:

- *Principal learning:* the mandatory 'core', specific to each Diploma, occupying around 40-50 per cent of a student's time;

- *Generic learning:* common to all 14 lines of learning, covering literacy, numeracy and IT, personal learning and thinking skills and an extended project (15-40 per cent of the student's time);

- *Additional/specialist learning:* further development of specialist skills relevant to the main thrust of the Diploma and/or a wider combination of options including one or more A-levels or GCSEs (20-30 per cent of the student's time);

- A minimum of 10 days' *work experience.*

Groups of employers came together to devise the principal learning component of the new Diplomas. Leaders of the engineering profession and manufacturing industries were especially enthusiastic: they saw an opportunity to challenge tired stereotypes about men (specifically men) in oily overalls, educate young people about the breadth, stretch and opportunities of engineering as a profession, and put them on track to be tomorrow's engineers and captains of industry.

Meanwhile, the government had also announced plans to introduce a 'Young Apprenticeship' (YA) scheme (Department for Education and Skills, 2004a). The first YA programme started in 2004. By 2010, up to 9000 students a year could choose from YAs in engineering; motor industry; business and administration; health and social care; art and design; performing arts; hospitality; electricity and power; hairdressing; construction; food and drink manufacturing; science; and sports management, leadership and coaching. Young Apprentices spent three days a week at school where they studied for GCSEs in English, maths, science and IT along with other national curriculum requirements such as religious education and physical education. The other two days were given over to work experience and studying for a vocational qualification at a further education college or other training provider.

While plans and policies were being drawn up at a national level, local schools and colleges had taken advantage of increased curriculum flexibility to offer a wider range of vocational opportunities, particularly in key stage 4. An Ofsted report suggested that local actions had re-engaged some young people, boosted motivation and improved behaviour:

> More appropriate curricula, particularly the provision of vocational courses, re-engaged many students. Behaviour and attendance improved and the courses raised the achievement of particular groups of students, particularly those at risk of disaffection or disengagement. (Ofsted, 2007, p.5)
>
> Teachers reported increased motivation and attendance where students studied new courses. The students were very positive about changes to the curriculum. Those in schools which had adapted their curriculum generally made better progress than similar cohorts in previous years, and in many cases than they had initially been expected to. (Ofsted, 2007, p.7)
>
> Over half of the schools visited attributed improved attendance to the availability of more appropriate courses for students ... Most schools had evidence that students who were at risk of becoming disaffected in Year 9 were enjoying learning more in Key Stage 4 because they could study subjects they felt were relevant and had more varied ways of learning. When inspectors spoke with students, they were able to confirm schools' views that, for many, being able to study things that interested them, and in a way that helped them learn and improved their self-confidence, motivated them to attend more regularly. Some parents spoke movingly of how schools had

helped change their children's approach to learning and taking control of their lives. (Ofsted, 2007, p.9)

The government believed that boosting vocational opportunities would increase achievement rates in key stage 4 and lead to a rise in the proportion of young people staying in education or training after the age of 16. At the same time, education ministers promised to create more academies. The next step would bring the threads together.

LORD BAKER AND LORD DEARING CALL FOR 14-19 SCHOOLS THROUGHOUT ENGLAND

In 2006, Lord Baker proposed setting up new 14-19 technical schools in towns and cities across England. Speaking in the House of Lords, he said:

> Two years ago, the Government stumbled on the correct policy and, following the Tomlinson report, decided to develop a 14 to 19 curriculum ... but institutions that can deliver the 14 to 19 curriculum are missing.
> Therefore, I advocate as a serious policy that the Government should establish as new secondary schools only schools for 14 to 19 year-olds. Call them what you will – skills schools or apprenticeship schools – but they are a new type of school. Tony Blair has already said that he is committed to 200 new academies as part of his legacy. I welcome that, but I wish that he would be more radical. It is no good him relaunching what I did 20 years ago. If he wants a good legacy, leaving 50 or 60 14 to 19 schools would be a legacy worth having.
> (HL Deb 13 December 2006, col. 1557)

Lord Baker's vision was shared by a friend and fellow member of the House of Lords, Lord Dearing.

Ron Dearing left school at 16 and joined the civil service as a clerk. He worked in the Ministry of Power before moving first to HM Treasury and later to the Department for Trade and Industry. After leaving the civil service, Sir Ron – as he was by then – was appointed chair of the Post Office. At the time, the minister responsible for the Post Office was Kenneth Baker and the two became firm friends. Kenneth Baker later invited Sir Ron to chair the Council for National Academic Awards (CNAA), the body responsible for validating polytechnic degrees.

In the 1990s, Sir Ron completed three reviews for the government, covering the national curriculum, qualifications for 16-19 year olds, and higher education. He was made a life peer in 1998.

In the early 2000s, Lord Dearing came to appreciate and support the Labour government's academies programme. He felt it could provide a route to establish the kind of 14-19 technical institutions which he and Lord Baker had in mind. Lord Baker said:

> I knew Ron Dearing very well. Ron left school at 16 and eventually became a senior civil servant and head of a nationalised industry – you could say he pulled himself up by his own bootstraps. He was the best chairman the Post Office ever had and workers loved him – they really did – and he loved the institution. I was so impressed by him that when he retired [from the civil service], I was one of the first to offer him a job, as chair of CNAA, the body dealing with qualifications awarded by polytechnics.
>
> I had of course kept in touch with Ron: we were both members of the House of Lords by then. Ron always had a big interest in technical education – he wrote a report on it in the mid-1990s. We put our heads together to devise the best way to tackle the problem. Ron had a lot of very helpful experience from working with the Church of England to establish academies. He said we should set up new technical schools as academies, because this would provide them with freedom, control over their budgets, and so on. We decided that at a very early stage – a sensible decision. We also decided that young people should join our technical schools at 14, not 11, and be offered education leading to very good career opportunities.
>
> We had to elevate the status of technical education. We both agreed the way to do it was to get universities

involved. I think it was a joint idea, but I was very enthusiastic about it.

Thirdly, with my business background and drawing on experience with CTCs, I said businesses needed to be involved more than ever before.

So we had three distinctive features: 14-18, university sponsorship, and active support from industry.

(Baker K., 2011c)

On 15 October 2007, Lord Baker told Andrew Adonis that he and Lord Dearing wanted to start a specialist 14-19 academy. Lord Adonis – who had been made a life peer and appointed Minister of State for Education in 2005 – was extremely interested and suggested opening two specialist academies of this kind.

Three weeks later, Lord Dearing spoke in another debate in the House of Lords:

> [To] address the skills agenda and to make more young people want to continue in learning – which I believe is an essential element in a successful policy to extend the learning age to 18, for coercion on any scale simply will not work – we have to rethink the comprehensive model of education for 14 to 18 year-olds. It is not the best model for delivering the Government's agenda for world-class skills on the scale required. We need to be able to offer young people the opportunity at age 14 to go to a college of technology in our cities and big towns specifically to develop skills preparatory to an apprenticeship or to pursue specialised diplomas that require specialist equipment…
>
> If we are going to extend the learning age to 18, as proposed, we will inevitably need new buildings. We shall need to enlarge our cadre of teachers skilled in teaching the skills that industry and commerce need. We would be seen to be taking skills very seriously if 14, 15, 16 and 17 year-olds, rather than having to move from one comprehensive school to another or to a [further education] college to do their specialist work, were able to go to purpose-built colleges of technology,

continually equipped with the latest tackle needed to develop their skills.
(HL Deb 8 November 2007, col. 221)

In January 2008, Lord Baker and Lord Dearing visited Rocester to hear about plans for the JCB Academy and in the following months, they toured the country meeting vice chancellors, college principals, employers and the other potential partners who might help open 14-19 technical schools.

Lord Baker and Lord Dearing were offered administrative support by the Edge Foundation, an independent education charity launched in 2004 to champion technical, practical and vocational forms of learning. Edge agreed to release one of their staff, Jane Samuels, for three days a week. Her role description was drawn up by Peter Mitchell, who joined the staff of the Edge Foundation after retiring from his position as head teacher of Thomas Alleyne's High School, Uttoxeter – a fitting connection, given his involvement in early planning for the JCB Academy.

In the summer of 2008, Baker and Dearing decided on a name which would distinguish their institutions from other academies. They chose 'University Technical College': 'university' because each would be sponsored by a university and provide a pathway to higher education for those students who wanted it; 'technical' to emphasise the specialist curriculum and links with employers; and 'college' to differentiate the new institutions from mainstream schools and academies. UTCs would respond directly to the government's priorities for secondary education and skills by combining a core curriculum with Diplomas and Young Apprenticeships, delivered in close partnership with universities and employers.

UTCs would recruit at 14 rather than 11. One reason was to align with the Labour government's concept of a 14-19 phase of education. Another was Lord Baker's belief that young people would be better placed to choose a technical pathway at 14 than they were at 11. He said:

> If I was devising the national curriculum today, I would stop it at 14. I think there is a very strong argument that all young people should go through what R A Butler [Minister of Education from 1941

to 1945] described as the mill of education, giving them a general education in English, maths, science, history, geography, art and music – a fully rounded education. But at 14, young people should decide for themselves what direction they want to take, provided they can move later if they have taken the wrong decision. I believe there should be four distinct pathways – the academic, grammar-style pathway; the technical pathway that we are pioneering in University Technical Colleges; a vocational pathway covering subjects which are less technical but even more hands-on; and a creative arts pathway covering music, art, drama and all of that.
(Baker K., 2011c)

5
TOWARDS THE 2010 GENERAL ELECTION

UTCs were designed to deliver some of the Labour government's most important educational aims. They would –

- be established as part of the academies programme,

- would respond directly to the concept of a 14-19 phase of education,

- provide a curriculum built around the new Diplomas and potentially Young Apprenticeships, and

- encourage young people to stay in education or training until they were 18, including some who might otherwise drop out early.

UTCs enjoyed the active support and involvement of employers, who co-designed and delivered the curriculum and promised other opportunities for UTC students such as work placements and routes into apprenticeships. But the concept would need support from leading politicians across the political divide, because a general election was looming.

Lord Baker was particularly keen to persuade his own Conservative Party colleagues to support the UTC concept. He prepared a paper which he sent to the then Leader of the Opposition, David Cameron, in August 2008. Lord Baker recalled:

> David [Cameron] agreed I could give a presentation to him, his advisor Steve Hilton, Oliver Letwin,

George Osborne, Michael Gove and David Willetts – they were all there and they all liked what they heard. I think David Cameron was particularly turned on by the idea that this would help with his idea of the broken society. UTCs would help engage and capture the enthusiasm of young people.
(Baker K., 2011c)

Lord Baker and Lord Dearing briefed parliamentarians about their ideas and spoke in the House of Lords whenever they had the opportunity. They both took part in a debate on the Queen's Speech on 11 December 2008. Lord Baker spoke first:

My old friend the noble Lord, Lord Dearing, and I have been promoting a new type of school. We call them university technical colleges. The idea is that they will be what they say they are – technical vocational schools – for 14 to 19 year-olds …

For those aged 14, there will be two sources of entry: one for [young] apprenticeships and one for students doing Diplomas. They are interchangeable. The universities are interested because they see this as a pathway to foundation degrees and even to higher degrees. I am glad to see that the Government fully support them; they certainly have the full support of the noble Lord, Lord Adonis, and his successor [as Minister of State for Schools], Jim Knight …

We should have had technical schools since 1870. We have tried. Butler tried. Butler's great plan [enabled by the 1944 Education Act] was for grammar schools, secondary modern schools and technical schools. The first ones to go were the technical schools – infra dig greasy rags. That was a huge mistake. The only country that adopted the Butler scheme was Germany, which has grammar schools, high schools and technical schools. A report published this year in Germany on the whole education system said that the most popular and successful schools in Germany are not the grammar schools – that must please some Members on the Benches opposite – but the technical

schools. It is about time that we had such schools here in our country.
(HL Deb 8 December 2008 cols. 512-3)

Lord Dearing added:

> The initiative that the noble Lord [Baker] and I have taken has been to try to ... engage not only the brain but the hand, the eye and the equipment by doing things in a much more hands-on way, because that will appeal to a wider range of people. The involvement of the universities will mean the highest quality and will engender great respect as well as commitment.
> (HL Deb 8 December 2008 col. 553)

Speaking from the Conservative front bench, Baroness Verma said, 'My noble friend Lord Baker of Dorking speaks with huge authority and expertise on education. He highlighted with passion the need for good quality delivery of vocational diplomas to meet the ever-changing demands of a competitive economy' (HL Deb 8 December 2008 col 573). Across the floor, Lord McKenzie – a junior minister in the Labour government – said, 'I applaud the undertakings of the noble Lords, Lord Dearing and Lord Baker, to establish university technology colleges' (HL Deb 8 December 2008 col. 580).

There was a clear cross-party view that UTCs could be a valuable part of the educational landscape. This was confirmed in the autumn of 2009.

Lord Baker was asked to speak at the Conservative Party conference. He was followed by the Shadow Secretary of State for Children, Schools and Families, Michael Gove, who said a future Conservative government would create new technical schools in major cities. A press release issued the same day said:

> *Gove: A new generation of Technical Schools*
> Conservatives are today pledging to build a new Technical School in each of the 12 biggest cities in England, with the long-term ambition to have one in every area of the country.
>
> These high quality, high-tech academies will raise

the status of technical qualifications, boost Britain's science and engineering base, and provide real choice for parents and young people.

The new Technical Schools will have academy status, be open to all students from 14 years of age and be sponsored by leading businesses and universities, working with a Trust set up by Lord Baker:

Our radical plans for school reform will support plans for Lord Baker's Trust to open Technical Schools in England's 12 largest cities, and in the long run we would like to see a technical school in every area.

Funding will be made available for these new schools from the existing academies budget.

The Conservatives will also lift the cap on young apprenticeships so that all 14-16 year olds have access to genuinely vocational qualifications. A Conservative Government will fund 30,000 places a year (compared with the present 10,000). Schools will be allowed to offer self-funded places if there is demand beyond 30,000 places.

Shadow Secretary of State for Children, Schools and Families, Michael Gove, said:

'Our new technical schools will provide credible, high quality vocational education in each major city. We will also triple the number of Young Apprenticeship places to 30,000 and remove the cap that stops state schools offering these places. This is crucial to tackling youth unemployment and recovering from the recession.'

(Conservative Party, 2009)

A month later, the Department for Business, Innovation and Skills published a skills strategy, which pledged cross-departmental support for UTCs:

We will work with the Department for Children, Schools and Families to support the development of University Technical Colleges. These will offer new opportunities for 14-19 year olds to undertake vocational and applied study. Alongside the introduction of 14-19 Diplomas, University Technical Colleges will greatly strengthen the flow of young people coming into the labour market with the skills and capabilities employers want and particularly for technician careers. We will ensure good progression from University Technical Colleges to other routes of study including advanced apprenticeships and foundation degrees.

(Department for Business, Innovation and Skills, 2009, p. 30)

Both parties included UTC pledges in their manifestos for the 2010 general election.

LORD DEARING

Lord Dearing died in February 2009. A service of thanksgiving was held at St Margaret's Church, Westminster Abbey. Lord Baker made an address, praising Ron Dearing's life, career and quiet passion for education. He said:

Ron particularly wanted to help the less advantaged. Both he and I saw great force in the phrase from one of Shakespeare's plays, 'the fire i' th' flint shows not till it be struck'. Even the pupil who is most turned-off and fed up with school has that piece of flint and the whole purpose of education is to find it and to strike a spark.

To achieve this he and I for the last 18 months had been working on a new type of school. We both believed that 14 is a more natural age of transfer than 11, and that the greatest loss to British education was the loss of the technical schools that Butler left behind in 1945. So, with the support of the government, we have persuaded several universities to sponsor new colleges – University Technical Colleges – for 14-19 year olds taking the new Diplomas in engineering, building, construction, manufacturing, production, design and IT, with two forms of entry: one for apprentices and one for students. The first UTC was approved in Birmingham in December [AUEA] and there are several more in the pipeline. These UTCs will be an enduring legacy.

Campaigning with Ron over the last few months was a wonderful experience and great fun. He did not 'go gently into that good night' – he spoke in the House of Lords just a few weeks before he died. Ron was the epitome of Victorian values – education was a ladder, hard work and application were the keys, helping others less fortunate a duty. His most endearing quality – and this is not something that can be learnt or is part of the curriculum – was enthusiasm. The sparkling brightness of his eyes, the sheer boyish glee of getting things done, of moving on, but guided all the time with great modesty, faith and compassion.

(Baker K., 2009)

THE BAKER DEARING EDUCATIONAL TRUST

Lord Dearing's contribution was further recognised when a new trust was established to promote and support the development of UTCs. The Baker Dearing Educational Trust was registered with the Charity Commission on 8 May 2009 'to advance the education of children, young people and young adults by technically oriented study at new or existing colleges' (Baker Dearing Educational Trust 2009a).

The Baker Dearing Educational Trust (usually abbreviated to BDT) was inspired in large part by the CTC Trust which helped develop and support City Technology Colleges. The idea was to create an independent charity which could liaise with the Department for Children, Schools and Families, universities and employers; encourage interest in establishing UTCs; advise local partnerships on expressions of interest; and provide advice to principals and sponsors once their UTCs were up and running.

Lord Baker was elected chair of BDT's board of trustees. Other early trustee appointments included Dame Ruth Silver, principal of Lewisham College until 2009 and subsequently chair of the Learning and Skills Improvement Service; Sir Mike Tomlinson, former Chief Inspector of Schools and lead author of the Tomlinson Report; Sir

John Rose, chief executive of Rolls-Royce; Allan Cook, chairman of W S Atkins, a multinational design, engineering and project management consultancy; Frank Field, Labour MP for Birkenhead; Lord Puttnam, former film producer; Nigel Thomas, representing the Gatsby Charitable Foundation, a significant financial supporter; and Gordon Birtwistle, elected to serve as Liberal Democrat MP for Burnley in 2010. Lord Adonis joined the board of trustees after the 2010 general election.

Sir Mike Tomlinson explained how he was recruited to the board:

> Kenneth Baker and I had run across each other a number of times. I remember one classic occasion when I was about to cross the road in Covent Garden when he drew alongside in his car, wound down the window and shouted, 'don't worry, we'll get on with it [the Tomlinson Report] sooner or later!' I also knew Ron very well – he had backed our reform ideas, too. Anyway, Lord Baker rang me and explained what he and Ron had been working on. He said, in a nutshell, 'we are trying to make a reality of your vision on technical education and 14-19, and we would like you to be a trustee'.
> (Tomlinson, 2017)

As interest in UTCs grew, BDT needed more capacity. The Edge Foundation offered support both in cash and in kind. Edge seconded Peter Mitchell to act as chief executive of BDT, supported by Lena Kokkolas (executive assistant) and Kathy Fogarty (company secretary). Additional support was provided part-time by Jane Ware, director of communications from January 2010 and later, director of programmes.

Charles Parker joined soon after as director of operations. He had been closely connected with the government's academies programme, first as clerk of the Mercers' Company – an early academy sponsor – and later as academies advisor with Partnerships for Schools, a non-departmental public body set up to manage investment in schools and academies. He later took over from Peter Mitchell as BDT's chief executive.

Lord Baker wanted UTC policy and practice to be informed by lessons from past initiatives and good practice around the world.

Edge agreed to commission a research project on BDT's behalf and in November 2009 accepted a proposal submitted by Professor William Richardson (University of Exeter) and Susanne Wiborg (Institute of Education) to provide (a) an account of the history of technical and vocational school-based education in the UK and (b) a comparative study of technical education in Germany, the USA, Japan and Sweden. Their report was called *English Technical and Vocational Education in Historical and Comparative Perspective* (Richardson and Wiborg, 2010).

In February 2010, BDT hosted a seminar on UTCs. Speakers included William Richardson, Paul Pritchard and Jim Wade from the JCB Academy, and prime movers in the establishment of the next two UTCs, Professor Alison Halstead, future chair of governors at Aston University Engineering Academy and Amarjit Basi, principal of Walsall College and leader of the team preparing to open Black Country UTC. These two UTCs are described in more detail in chapter 7, below.

Lord Baker recalled a pivotal moment during the seminar:

> We asked William Richardson to come and talk about his research report. He said at the end of his presentation, 'what you have to decide is whether this is going to be an experiment or a movement.' I realized that that was the crossing point. Twelve UTCs would be an experiment: a movement would be 50, 100 or 200. That was the moment we became much more directed to creating a movement.
> (Baker K., 2011)

6
THE COALITION GOVERNMENT

The general election took place on 6 May 2010. The Conservative Party gained the largest number of seats but fell short of an overall majority. A few days later, the Conservatives and Liberal Democrats agreed to form a coalition government. Michael Gove was appointed Secretary of State for Education.

Little was said about technical and vocational education in schools until September 2010, when Michael Gove delivered the Edge Foundation's inaugural annual lecture, hosted by its recently-appointed chair, Lord Baker.

Gove strongly criticized the previous Labour government's record on technical and vocational education in schools. He noted that in 2004, 22,500 vocational qualifications were taken in schools; five years later, the figure had risen to 540,000, an increase of 2,300 per cent. He questioned both the quality of the qualifications and the reasons why schools had enthusiastically adopted them:

> Some of these qualifications badged as vocational enjoy a ranking in league tables worth two or more GCSEs, making them attractive to schools anxious to boost their league table rankings. And that has meant that some schools have been tempted to steer students towards certain qualifications because it appears to be in the school's interests even when it's not in the student's.
> (Department for Education, 2010)

Gove said that a marked improvement in school-based vocational education could be achieved by boosting the quality of courses and qualifications on the one hand, and by creating new institutions on the other – including UTCs:

> We have to have courses, qualifications and institutions during the period of compulsory schooling which appeal to those whose aptitudes and ambitions incline them towards practical and technical learning.
>
> We're already using our radical schools reform programme to promote new institutions designed to support high-prestige technical education with a clear link to employment and further study.
>
> The university technical colleges – a model developed by my great reforming predecessor Lord Baker and the late Lord Dearing – tick all the boxes.
>
> The idea is very straightforward: technical colleges will offer high-quality technical qualifications in shortage subjects like engineering. They will do so as autonomous institutions – legally they will be academies – sponsored by at least one leading local business and a local university.
>
> The pattern for their success has already been set by the new JCB Academy in Staffordshire, which I was privileged to be able to visit earlier this year. It combines hard practical learning – with courses in technical subjects involving applied work of the most rigorous kind – alongside a series of academic GCSEs – including maths, English, science and a foreign language.
> (Department for Education, 2010)

Gove also spoke about his plans for an English Baccalaureate, which he described as 'a new certificate for all children who achieve a good GCSE pass in English, maths, a science, a modern or ancient language and a humanity like history or geography' (Department for Education and Skills, 2010). The English Baccalaureate – or EBacc for short – would also act as a new league table measure to encourage schools to give all young people a broad and rounded base of knowledge.

In his Edge lecture, Gove noted that other countries offered combined academic and technical programmes in schools. In the Netherlands, for example, all 16-year-olds were assessed in foreign languages, arts, sciences, maths and history, including children who move onto a technical route at twelve. In England, therefore, it would be possible to offer the English Baccalaureate alongside technical qualifications:

> Securing this core base of knowledge would not preclude the study of technical or vocational subjects as some have suggested. It's not either/or but both/and. I'm absolutely clear that every child should have the option of beginning study for a craft or trade from the age of 14 but that this should by complemented by a base of core academic knowledge.
>
> And the new generation of university technical colleges – by taking students from other schools at the age of 14 – will help secure this route. When we open a new UTC in Aston in 2012 pupils will specialise in engineering and manufacturing alongside core academic GCSE subjects. Crucially, students will have the opportunity to work with Aston University engineering staff and students as well as local businesses and further education colleges.
> (Department for Education, 2010)

In the weeks following the Edge lecture, the Secretary of State decided to grant limited funding to the Baker Dearing Educational Trust to support work with an ever-growing number of partners interested in opening UTCs. On 17 January 2011, he offered a grant of £200,000 for a 12 month period, to be followed by a full invitation to tender later in the year.

Shortly afterwards, Lord Baker hosted a dinner for vice-chancellors and business leaders interested in opening UTCs. Michael Gove and the skills minister, John Hayes, both attended. Baker made a note in his personal diary:

> John made a warmly supportive speech saying that Britain should be "peppered" with UTCs. Gove [also] spoke very warmly of UTCs and wants to see many more across the country: they are revolutionary and all that...
> (Baker, K., 2011a)

THE WOLF REPORT ON VOCATIONAL EDUCATION

Michael Gove took the opportunity of the Edge lecture to announce a review of vocational education:

> I'm absolutely delighted today to be able to announce that Alison Wolf – the Sir Roy Griffiths Professor of Public Sector Management at King's College London – has agreed to lead a review into pre-19 vocational education ... This review will be very different from previous efforts. It is not going to lead to yet another set of unwieldy, Whitehall-designed and short-lived qualifications, or a new set of curriculum quangos. Instead, we want to establish principles, and institutional arrangements, which will encourage flexibility and innovation.
> (Department for Education, 2010)

The Wolf Report on Vocational Education (Department for Education, 2011a) was published on 3 March 2011. Professor Wolf's central thesis was that both before and after the age of 16, too many young people were taking qualifications that were of little value in the labour market, while too few achieved a good standard in English and maths by the age of 18. In future, only 'rigorous' vocational qualifications should count in school league tables.

Professor Wolf supported a broad and balanced key stage 4 curriculum. While academic subjects should be at the heart of the curriculum, young people should be able to spend up to 20 per cent of their time studying practical and vocational subjects, provided they were taught by people with appropriate experience and expertise (including further education teachers), in well-equipped facilities. She said:

> ...colleges will generally be far better placed to provide vocational options for 14-16 year olds, and not just 16-18 year olds, than schools will. This has been the rationale behind special grant-bearing programmes such as Young Apprenticeship and Increased Flexibilities. But these are very expensive, benefit only a tiny, more or less random sub-set of the age group, and cannot, by virtue of their project funding and extra cost, become permanent and system-wide ...
>
> If colleges enrol students under 16 then they can revive junior technical provision ... Colleges must, however, also offer students a full 14-16 programme – or arrange collaborative teaching with local schools.
> (Department for Education, 2011a, p.128)

Professor Wolf also mentioned UTCs and another new category of 14-19 school, studio schools:

> ... well-equipped and specialist schools such as studio schools and University Technical Colleges will and should offer programmes with a strong and distinctive vocational element. They have the extra dedicated resources, and, in the case of UTCs, expect to operate with longer teaching days and years, making extra time available for specialist options.
> (Department for Education, 2011a, p. 111)

The Wolf Report was widely welcomed. A leader in *The Times* said, 'Taken in conjunction with the new movement of University Technical Colleges established by Lord Baker of Dorking and the late Ron Dearing, the Wolf Report offers the prospect of serious progress on an issue that is profoundly important for Britain's economic health' (Anon, 2011).

BECOMING A MOVEMENT

The new government accepted the Conservative Party's manifesto commitment to open 12 technical schools. With the Baker Dearing Educational Trust in place, regular meetings between BDT and Department for Education (DfE) officials started soon after the election. As early as August 2010, the BDT and DfE officials discussed progress towards opening Aston University Engineering Academy and Black Country UTC (see chapter 7, below), as well as the prospects for UTCs in over 25 other towns and cities. The programme was gaining momentum.

The Times gave UTCs top billing on Friday, 7 January 2011. Under the headline, 'Selection at 14 will drive revolution in schooling', Greg Hurst wrote about Baker's ambition to open up to 70 UTCs, attended by tens of thousands of pupils. He quoted Lord Baker as saying: 'This is a movement – not just a few experimental schools' (Hurst, 2011).

Elsewhere in the same edition, Laura Dixon reported on a visit to the JCB Academy, which opened the previous autumn. Describing the facilities as 'incredible', she wrote: 'Housed in the 18th century Arkwright Mill, which was producing cotton until the early 1990s, the old equipment has been replaced with lathes to turn metal, plastic moulds, plasma cutters, and 3-D screening rooms' (Dixon, 2011) .

The Times also published a comment by Professor Alan Smithers, Director of the Centre for Education and Employment Research at the University of Buckingham, who pointed out that many countries in Western Europe had some form of assessment that enabled young people to choose between different pathways. 'It's a matter of choice rather than selection,' he said. 'It's holding a mirror up to themselves so that they can see how they are doing. Fourteen is the right age for this' (Smithers, 2011).

The next day, *The Times* carried an opinion piece by Lord Adonis. He started with a declaration: one of the greatest disasters of post-war education policy in England was the failure to establish more than a few technical secondary schools. 'Now, at last,' he wrote, 'technical schools for 14 to 19-year olds are being created on a systematic basis, thanks to a brilliant initiative by Kenneth Baker and the late Lord Dearing' (Adonis, 2011).

That was all very encouraging, but UTCs could not go from experiment to movement without further investment from the public purse. Michael Gove spoke to Lord Baker about this on 10 March 2011. Lord Baker said:

> Basically he asked whether I would renew my pressure on and representations to the Chancellor to get more money. He knows he has got little available for UTCs after the first 12 and that many hopes and expectations are going to be dampened ... After the call I immediately drafted a letter to the Chancellor and made sure that Steve Hilton at No 10 saw it as well as Michael Gove.
> (Baker, K., 2011b)

A week later, the Chancellor of the Exchequer, George Osborne, called. Lord Baker recalled the conversation in his personal diary:

> I was asked to stand by to take a call from the Chancellor. When George rang he said he was really impressed with UTCs and wanted to help. Then he explored with me the cost. I said they cost about £8 million each [to build and equip] and if he wanted to double the programme then we are talking at least £100 million. Clearly he was sounding me out and I told him that if he allocated us funds I would be able to deliver an increased number of UTCs.
>
> This is tremendous news and clearly the pincer movement by Michael Gove and myself has produced a result. Even in these desperate times the Prime Minister and the Chancellor want to take initiatives which stimulate growth.
> (Baker, K., 2011c)

The Chancellor delivered his Budget statement on 23 March 2011. About half way through his speech, he said:

> The Government are committed to funding new university technical colleges, which will provide 11 to 19-year-olds with vocational training that is among the best in the world. The curriculum is being developed in close co-ordination with both local universities and

> leading employers. I commend Ken Baker on getting the new colleges up and running in our manufacturing heartlands. To date, the Government have announced that we will fund 12 new university technical colleges. I can tell the House that we will provide funding to double that number to at least 24.
> (HC Deb 23 March 2011, col. 960)

(The reference to '11 to 19'-year-olds was a slip of the tongue: UTCs were, at that time, targeted at the 14-19 age group.)

Lord Baker was delighted:

> I went along to listen to the budget statement. George [Osborne] saw me sitting in the gallery and when he got to the passage about UTCs, he looked up and mentioned me by name – which was most extraordinary! I was delighted that the enthusiasm we had managed to generate made that happen.
>
> The initial pledge for 24 UTCs means spending £200 million. I want to move on to the stage where we can do 50 or 100. I think we're lucky, because we have struck a chord at the right time. There is a readiness: people are ringing up and saying they want to talk to us. When I was in Plymouth last week, one of the industrialists had been up to see the JCB Academy: he thought it was magnificent and now he wants to do the same.
> (Baker, K., 2011c)

The news made the front page of that week's *Times Educational Supplement*, under the headline 'Osborne signals technical college surge'(Vaughan, 2011). The article quoted comments made by Stephen Capper, head of Sawyers Hall College in Essex, who described the announcement as a step towards a 'complete overhaul' of the education system. Elsewhere, the General Secretary of the National Union of Teachers, Christine Blower, said it was 'unacceptable to force pupils into specific learning routes at such an early age [as 14]', and Sally Hunt, General Secretary of the University and College Union, feared that UTCs would 'divert money away from further education colleges, reintroduce selection at 14 and create a two-tier system' (*Manchester Evening News*, 2011).

YOUNG APPRENTICESHIPS ARE ABOLISHED AND THE WOLF REPORT IS ACCEPTED

A few days after the Budget statement, the government withdrew support for Young Apprenticeships (YAs). An official in the Young People's Learning Agency announced the government's decision to end YAs in a note to local authorities. He said the decision had been made in the context of the DfE spending review settlement and that:

> The high cost of the programme could not be maintained in the current economic climate. It was also felt that that the programme would need to be significantly re-designed to fit with the English Baccalaureate and the recommendations of Professor Alison Wolf's Review of Vocational Education.
> (Wilson, J., 2011)

In July, the government accepted the Wolf Report in full. In particular, ministers promised to 'break free from the old equivalency based [school] performance tables and include only a set of clearly defined vocational qualifications which have the greatest benefit for the [14-16] age group.' (Department for Education, 2011b, p.5)

In fact, the government did not abolish 'equivalency based performance tables': they simply revised the definition of 'equivalency'. Qualifications previously treated as equivalent to two or more GCSEs – Diplomas, for example – were now deemed equivalent to just one GCSE. In addition, no more than two non-GCSEs would count towards a school's performance measure. If students took three non-GCSEs, one of them would not count in the league tables (summarized from Department for Education, 2011c).

This was, for all intents and purposes, the final nail in the Diploma's coffin. Funding had already been curtailed, and now Diplomas would count as no more than one GCSE in the league tables. Most schools responded by dropping large vocational qualifications, including Diplomas, and replacing them with a mixture of GCSEs and vocational qualifications which took no longer to teach than a single GCSE. Awarding organizations responded by developing new qualifications which met this new standard.

UTCs needed to find a way though the new qualifications landscape. BDT and the Edge Foundation asked the Royal Academy of Engineering (RAEng) to identify technical qualifications in science, technology, engineering and mathematics (STEM) that would be respected by the STEM community:

- alongside GCSEs, as the technical component of the UTC curriculum at Level 2 (key stage 4)
- in combinations, as the core of the UTC curriculum at Level 3 (post-16)
- individually and in combinations in other schools and colleges.

The resulting report, *Respected: Technical Qualifications Selected for Use in University Technical Colleges*, was published in October 2011 (Harrison, 2011). The author, Professor Matthew Harrison, established a robust method for reviewing individual STEM qualifications, taking account of factors including indicators of STEM community respect, current levels of take-up, and ethos – for example, whether qualifications encouraged and supported practical, hands-on learning. The report included relatively short lists of level 2 and 3 qualifications recommended for consideration by UTCs, which could be added to as awarding organizations developed the next generation of qualifications.

7

ASTON UNIVERSITY ENGINEERING ACADEMY AND BLACK COUNTRY UNIVERSITY TECHNICAL COLLEGE

ASTON UNIVERSITY ENGINEERING ACADEMY

Plans for a UTC in Aston began when Lord Baker, Lord Dearing and Jane Samuels met senior staff from Aston University and City of Birmingham Council on 28 April 2008. Lord Baker recalled:

> I rang Julia King, Vice Chancellor at Aston University. I gave her the spiel and she agreed to see us. Julia brought in her Pro-Vice Chancellor, Alison Halstead, who became a real driving force. She advised us to speak to the leader of Birmingham City Council. He brought in the cabinet member for education and his Director of Children's Services.
>
> Incidentally, I was the one to ring vice chancellors, because if you are a former Secretary of State for Education, vice chancellors will at least take your call!
> (Baker K., 2011c)

Lord Baker started by outlining the vision for a technical college for the 14-19 age group. It would, he said, be created as part of the academies programme and have an engineering specialism. It would offer a core curriculum including English, maths, science, ICT and religious education alongside the new Diplomas in engineering and Young Apprenticeships. Other options might include modern foreign languages relevant to engineering and international business – German, Hindi or Mandarin, perhaps – and business studies. The school day and the school year would both be longer than in conventional secondary schools. Critically, the new academy must be actively supported by employers.

Aston University and Birmingham City Council agreed to prepare a formal expression of interest with support from a consultant nominated by the Department for Children, Schools and Families (DCSF), which had replaced the Department for Education and Skills in 2007.

Two days after their meeting in Birmingham, the two Lords spoke to officials in the academies team at DCSF and the following day, they met the Director General of the City and Guilds of London to discuss qualifications. Things were starting to move very quickly.

Over the following weeks, Aston's expression of interest (EoI) went through a number of iterations. Initially, it included a proposal to select applicants on the basis of aptitude in engineering and related disciplines. The DCSF ruled this out for students applying to enter at 14: normal, non-selective admissions policies would have to apply. Other contentious issues included the potential impact on existing schools and other providers and even the name of the new school: should it be called a 'college' or an 'academy'?

The curriculum aims were less controversial. At key stage 4, students would study English, maths, science, information and communications technology, religious education, citizenship and physical education. This would account for 25 taught hours per week. The remainder of the timetable would be devoted to one of the technical specialisms: engineering and manufacturing, and business and enterprise. In addition, all students would develop skills for employment including communication, literacy, numeracy, creativity, collaboration, teamwork and leadership. Post-16 students would spend three and a half days a week on their chosen technical specialism; they would also choose from additional subjects including maths, ICT, languages and business. Across the whole age range, there would be an emphasis on both independent learning and individual and group projects. Students would be well prepared for employment, though many might go on to higher education at the end of year 13.

The EoI emphasised Aston University's strong links with employers, including both large businesses (National Grid, Rolls-Royce and Tata Industries, for example) and small and medium-sized enterprises across the West Midlands.

The EoI said that the impact on nearby schools would be minimal: students would be recruited from the whole of Birmingham and beyond, with an average of two students transferring from each of about 75 different schools.

The DCSF approved the final EoI in 2009. The next step was to prepare a full feasibility study, including plans for purpose-built premises on the Birmingham Science Park. The school was called the Aston University Engineering Academy (AUEA). Speaking in 2011, Alison Halstead explained the university's keen interest in establishing the AUEA:

> The vision which Kenneth and the late Ron Dearing shared with Aston University was about the importance of young people seeing that education is of the highest quality and can help them get to university. But it's not solely about going to university: we need to work with employers and further education to offer other pathways to work and success.
>
> Why would a university get engaged? If you are visited by people with the passion of Kenneth and Ron, that goes a long way! But that doesn't satisfy a governing body. A university will only get involved if it resonates with their mission and vision. At Aston, we have a strong commitment to engineering and links with industry. All our [undergraduate] programmes include a placement year. We have one of the very highest graduate employment rates in the country – this year, we beat Oxford, Cambridge and Durham!
>
> But we're also very good at widening participation. 57% of our young people are from the West Midlands region and 22% are from the lowest socio-economic backgrounds, with a family income of less than £18,000. That is a flavour of what Aston stands for.
>
> Against that background, we saw the opportunity

to bring our employers to the table, together with all the work we had done with schools already, in the setting of a UTC.

Staff in HE are under great pressure to get into schools to raise aspirations – but we tend to do it a little randomly. The UTC gives us clear structure. Our staff can input into curriculum development and share their knowledge, working directly with employers, recent graduates, teachers and students. They will be able to provide an answer to the question, 'Why am I learning this?'

(Halstead, 2011)

AUEA opened in 2012 with 54 students in year 10 and 107 in year 12. AUEA's motto was 'engineering and science excellence for 14-19 year olds in Birmingham'. All year 10 students learned practical skills by using manufacturing machinery and bench-fitting tools. Year 10 and 12 students visited businesses including National Grid, BMW and Cadbury. In enrichment sessions, they could take part in competitions such as F1 in Schools and VEX Robotics, or join clubs – the electronics club, for example, supported by Aston University STEM ambassadors. There was a strong focus on the practical application of science and hands-on laboratory skills, as well as opportunities to create computer games. From day one, AUEA emphasised careers education and provided insights into higher education and apprenticeships.

AUEA started its second year with about 300 students on roll. That said, a number of schools seemed to be encouraging some of their more challenging students to apply for places at the Academy, rather than students with a strong interest in the specialist engineering and science curriculum. This soon became a theme across the UTC movement.

Ofsted inspected AUEA on 4 and 5 June 2014, finding it to be 'good' in all categories (Ofsted 2014d). Teachers' expectations were high, lessons were strongly linked to the activities typically encountered in the workplace and students had access to excellent facilities and equipment. The proportion of students from minority ethnic groups and the proportion who spoke English as an additional language were both well above the national average.

Similarly, an above-average proportion of students were eligible for free school meals.

Ofsted commented favourably on links with employers. For example:

> In the sixth form teaching is enhanced greatly through strong links to problems associated with real life; for instance in an engineering lesson students took responsibility for merging what they had learnt in physics, electrical and mechanical engineering to produce working solar cells in response to a brief from an energy company that is linked to the academy.
> (Ofsted 2014d, p.5)

BLACK COUNTRY UNIVERSITY TECHNICAL COLLEGE

Amarjit Basi, Principal and CEO of Walsall College, was convinced that 'the time is ripe for the refreshing of local industry and for exciting young people about the opportunities that a modern manufacturing sector has to offer' (Gray, 2010).

Walsall College agreed to act as lead sponsor of a UTC in the Black Country, with Wolverhampton University as co-sponsor. The curriculum would be based around engineering and science, with a particular emphasis on 'clean engineering'. The UTC would offer Diplomas in engineering, manufacturing and IT as well as core GCSEs in maths, English and science.

In terms of premises, Sneyd Community School in Bloxwich was due to close. Although far larger than the UTC would need, the school buildings came equipped with essential facilities such as school kitchens and playing fields. At the time, accepting the site in Bloxwich seemed straightforward, logical and affordable. It also determined the start date: Sneyd Community School closed at the end of the summer term 2011, and Black Country UTC moved in straight away. As a result, it opened a year earlier than AUEA.

The UTC's principal, Chris Hilton, spoke at an event organized by the Baker Dearing Educational Trust in November 2011:

I've got the task of communicating the real excitement of the first term of our UTC. We are going back to the idea of a real technical specialism, of equipping young people for the future – that's the excitement I have.

We have a curriculum which covers the basics of English, maths, science and languages. Alongside that, we offer the challenge of working with Young Engineers and Young Enterprise, of making rockets and taking part in enterprise challenges. The Siemens Sleuth Challenge was a great example, right away in our first term.

The fact that we've put together a new staff team including engineers – people from industry and further and higher education – is exciting as well.

The project-based curriculum we are working on is an amazing concept. We are teaching subjects in an integrated way, building English, maths and other subjects into engineering. One of my engineers started a lesson in German, which is the way we intend to teach. It was a pleasure to see that happen.

We work with firms like Siemens, South Staffs Water and Haughton Design, a local design company which has worked with year 10 students to design an all-in-one travel toothbrush. You twist the handle and toothpaste is spread onto the brush – it's real design, a real, practical activity. It's exciting to see inventive young minds working on the project, working without boundaries. They will take this passion on into the engineering Diploma.

Running the school day on business lines, starting at 8-30 and ending at 5 (except on Fridays, when we finish at 4) is a challenge, but it's an exciting challenge. It gives us the chance to build in other activities – we play football every Thursday, for example, and all sorts of other enrichment activities.

(Hilton, 2011)

Amarjit Basi summed up the aims and purpose of Black Country UTC in a note for potential industry partners:

> The Black Country UTC will enable young people aspiring towards a career in engineering and science to benefit from a focused curriculum shaped by employers, delivered within an inspirational environment, supported by expert staff, and utilising the latest technology that industry has to offer ...
>
> The national concern with youth disengagement has identified high levels of truancy and NEETS (not in employment, education and training) alongside a general malaise amongst young people, unsure of employment prospects, as being particularly damaging to the regeneration prospects of regions such as the Black Country which the UTC will serve.
>
> (Black Country UTC, 2010)

Walsall College was lead sponsor, with the University of Wolverhampton acting as co-sponsor. The UTC would specialise in engineering and science, and in particular process engineering, product design and environmental sustainability. The UTC would provide students with:

- A specialist vocational pathway linked to traditional GCSEs, AS and A2 awards and the new Higher and Advanced Diplomas, plus opportunities to progress to Level 4 (Foundation Degree)
- A vocational pathway linked to BTEC Diplomas and National Vocational Qualifications
- Apprenticeships
- Enrolment with the Institute of Mechanical Engineers providing professional registration and the nomenclature of 'Young Engineer'

(Black Country UTC, 2010)

Both the school day and the school year would be longer than usual. Diploma learning, GCSEs and project work would take place during four terms, each lasting eight weeks. A fifth term would be used for enrichment activities, accelerated learning, extended project work, work-related learning, international exchanges, school visits and personal project work.

In the place of traditional house systems, the UTC created 'learning companies' to support student development and welfare and facilitate project activities to promote employability skills including communication, teamwork and enterprise.

The appeal for support from industry was successful. The multinational engineering business, Siemens, came on board in January 2011. The company provided all the students with uniforms and donated specialist equipment. In addition, Siemens engineers helped train UTC staff in the use of the equipment and regularly visited the UTC to discuss engineering with the students. Help was also promised by around forty other businesses, ranging from small, local companies to major employers including National Grid, Caterpillar, Chamberlin plc, ZF Lemförder, Sandvik, Stratasys and South Staffordshire Water.

Meanwhile, the UTC had recruited the principal, Chris Hilton – previously head of Kingsmead Technology School in Staffordshire.

All seemed to be well when the Secretary of State, Michael Gove, signed the UTC's funding agreement in May 2011. Part of the former Sneyd Community School was rapidly refurbished and equipped with engineering workshops, six science laboratories, a learning resource area and specialist ICT facilities.

However, the biggest challenge still lay ahead: recruiting students. Local schools did not pass information about the UTC to parents or students, which meant recruitment depended heavily on a combination of word of mouth, newspaper and radio advertising, mailshots, and events at Walsall College, the University of Wolverhampton, shopping centres and community venues.

By the time Black Country UTC opened in September 2011, only 120 students had enrolled, split unevenly between years 10 and 12. On the plus side, over a third of the year 10 students were female; on the minus side, overall numbers were significantly below expectations. Although some faced a long journey between home and the UTC, a significant number had previously been students at Sneyd Community School and enrolled simply because they lived nearby, not because they were especially interested in the specialist UTC curriculum.

To make good use of the spare capacity, Walsall College offered engineering apprenticeships in partnership with the UTC.

Apprentices attended the UTC to study for technical certificates and use the UTC's specialist equipment.

Working with industry partners, the UTC devised a range of cross-curricular projects for key stage 4 students, working with employers including:

- Haughton Design – design an entertainment system for children and young adults for use in a hospital environment
- Sandvik – manufacture an accurate component from the system designed for the Haughton Design project
- ZF Lemförder – produce 5 finished components using computer numerical controlled (CNC) machining operations
- Siemens – design and produce a system for the automation of heating and lighting control within a given building

Post-16 employer-linked projects included:

- Stratasys – suggest product improvements through the use of rapid prototyping, additive manufacturing and thermoplastics
- Finning – carry out diagnostic maintenance on a caterpillar excavator to include oil analysis, mean time to failure rates and detailed measurement
- South Staffs Water – assess the impact of engineering on the environment including analysis of water, air and soil
- Chamberlin – produce components using casting processes with explanations of the structural change to the material

Employers also sponsored individual units of Diploma courses and the University of Wolverhampton provided access to a range of facilities and offered tutor-led sessions in the university's engineering workshops.

There were high hopes that more young people would choose the UTC in 2012. In practice, enrolments into year 10 fell, despite a professionally managed marketing campaign funded in part by Siemens. In the 2012-13 academic year there were 63 students in key stage 4 and 112 in the sixth form.

Furthermore, there was anecdotal evidence that some year 10 students had been encouraged to leave their previous schools for perhaps the wrong reasons. Some had a history of poor attendance and behaviour. A number of students joined the UTC from pupil referral units after being permanently excluded by their previous schools. Staff at the UTC worked hard to meet their needs, including providing additional support for students with special educational needs and disabilities. Alex Reynolds, Black Country UTC's director of engineering (and later, principal of UTC Sheffield City Centre), spoke about this with pride:

> In the year 10 cohort at Black Country UTC, 50 per cent were challenging. But lots of those students went on and did some great things. One of the interesting things, and UTCs don't really get credit for it – in some ways, schools in general don't get credit for it – there were students in that cohort at Black Country who would have ended up involved in drugs or crime, or all sorts of things. It took us a long time, but we did change some of those students for the better. Maybe they didn't leave with amazing GCSE results, but they left at 16 having completed secondary education and moved into a job or further education with a sense of self-worth. And at the same time, some students left with amazing sets of GCSE results and went on to do amazing things.
> (Reynolds, 2018)

The first independent assessment of the UTC's performance came when Ofsted carried out an inspection in January 2013. Inspectors noted that some of the UTC's students had reputations for poor attendance and/or behaviour at their previous schools. As their report put it, 'A significant minority of students have gaps in their learning as a result of disrupted educational experiences in previous schools' (Ofsted, 2013, p. 3).

In the main, Ofsted's inspectors felt the UTC had risen to the challenge:

> Below average attendance, above average persistent absence and the high level of fixed-term exclusions are characteristics of the earlier school experience of many of the students. Now, at the [UTC], almost all of the most vulnerable students are becoming better behaved and their attendance is improving. This is

> helped by good support from Walsall Education Welfare Service staff and the effective work of the college's own learning mentor ... There is also very good support for students who are re-integrating back into mainstream schooling from pupil referral units. Their achievement, attendance and behaviour are steadily improving.
> (Ofsted, 2013, p. 5)

On the other hand, the curriculum did not appeal to all students equally. Ofsted found that while most students were eager to settle and get on with their work, a few lost concentration when they found the work uninteresting. This could sometimes spiral into disruptive behaviour which interrupted the whole class.

More broadly, Ofsted found that some teaching in key stage 4 required improvement because it did not sufficiently engage, challenge or enthuse students. As a result, progress was below expectations in some subjects. On the other hand, standards were above average in engineering, ICT and English, and sixth form teaching was consistently good:

> In good lessons, teachers and other staff lead the students' learning but never do the work for them. For example, in a science lesson, every student had to attempt a 'dusting' of a forensic piece of evidence and carry out detailed measures of other evidence. An engineering lesson expected students to use technically very demanding computer aided design (CAD) software. Yet each student had a go and tried again if things did not work and was coached to discover the right thing to do through the teacher's skilful questioning.
> (Ofsted 2013, p. 4-5)

Leadership and management were found to be good, too: 'Governors, school leaders and staff are managing the rapid development of this innovative engineering college well [and] leaders are tackling the weaker areas' (Ofsted 2013, p. 1).

Weighing all the strengths and weaknesses in the balance, Ofsted's overall conclusion was that the UTC required improvement. Steps taken in response to the report included engaging teachers from two local, high-performing schools to provide intensive support to selected year 11 students as they prepared for GCSE exams in English, maths and science.

A ROYAL VISIT

HRH The Duke of York visited Black Country UTC in July 2013. Speaking to an audience of students, staff and partners, he said:

> UTCs are about much more than GCSEs and A-levels alone. They provide an education on a far broader canvas, not only in engineering but also in the fields of employability and connectivity. All this is brought about through businesses – and engineering businesses in particular – engaging with education.
> (Duke of York, 2013)

Other speakers included students and employers. Joe Symonds was in year 13 at the time of the royal visit. He had originally planned to take A-levels in English literature, film studies and psychology. However, when he heard about Black Country UTC he appreciated it would offer something different and in many ways, more challenging. He said:

> To get recognised by companies in an increasingly competitive job market, I knew that qualifications alone would not be enough. At the UTC, I've been able to work on assignments where the brief was set by major engineering companies such as Siemens rather than teachers alone.
> I've also spent time in businesses where I've performed maintenance checks, collected destructive and non-destructive test data and learnt about injection moulding at first hand.
> (Symonds, 2013)

Mwaka Musuumba described her experiences as a year 11 student:

During my first two years at Black Country UTC I have been involved with a number of projects and companies. One example was designing a hanging basket bracket as part of a project set by Haughton Design. We were encouraged to be innovative and create something different. I decided to make a bracket based on the 2012 Olympics. After the Games finished, the bracket could be redesigned with the 2014 World Cup in mind. The project had three phases: design, prototype and presentation to Haughton Design.

Another project was set by ZF Lemförder. We had to work in a team to create ball joints for Aston Martins, using a CNC machine. Working together professionally as a group, we were able to manufacture the parts within the required tolerances and specifications.

The way projects are set helps us develop teamwork, leadership and communication skills. We also meet people from different professions and learn about their areas of expertise. This has helped me realize what I want to do in the near future, which is to work in renewable energies. My ambition is to fight against climate change and help make the world a better place. By staying at the UTC I believe I will be able to achieve my dreams.
(Musuumba, 2013)

David Mills, managing director of Haughton Design, explained why he had decided to support the UTC:

My company offers a complete design and development service for large international companies, SMEs and inventors across a range of sectors including aerospace, defence and medical devices. We currently employ eight people.

I am a very practical person and struggled to apply myself at school – what I really needed was to get out from behind the desk and do something more practical. I'm pleased to say that I went on to become an engineer and then started my own company.

The reason I got involved with Black Country UTC

is that the country – and my clients – need people with the skills to make things, and to make things happen. With help from companies like mine, more young people will be able to go on to engineering and manufacturing – and maybe even start their own businesses in the future.

There's not been a great deal of effort required on my part. When the product design module was launched, we spent half a day at the UTC introducing the challenge to young people – and it was a pleasurable experience, something we're not used to doing. After we set the module, students worked on their ideas, supported by brilliant facilities and teachers. We came back a couple of months later for a dragon's den-style presentation. It must have been really quite intimidating for the students, but we heard seven really strong presentations. I was stunned that after only three months here, students could produce work of such high quality.

Subsequently, one of the students in the winning team, George, came to Haughton Design on work experience. He's so keen that he'll be working for us again during the summer holiday.

From my experience even very small businesses can make a difference, and at the same time put something back into education. I've met some fantastic young people. I've also met some great companies through my connections with the UTC. Most of all, I feel rewarded because I am making a difference to young people's education and future careers.

(Mills, 2013)

Haughton Design employed just eight people in 2013. At the other end of the spectrum, Siemens operated worldwide. Brian Holliday, Divisional Director for Industry Automation, was the company's main link with Black Country UTC. He said:

I am delighted we have been able to bring some of our technology here, and to help students design projects, solve problems and tackle engineering challenges.

Engineering is important to many of the megatrends affecting our planet, such as urbanisation – more people living in cities; demographic change – people living longer; globalisation – more competitive pressure on manufacturing; and, of course, climate change.
(Holliday, 2013)

8

THE NEXT WAVE OF UTCs

By July 2011, the JCB Academy was completing its first academic year, the Black Country UTC was preparing to open, construction of the Aston University Engineering Academy had started and agreement had been reached to open UTCs in Hackney and Greenwich. Applications had been considered more or less on a case by case basis. Now that there was support for 24 UTCs, interest had mushroomed: the Department for Education received 51 statements of intent to open UTCs in the first half of 2011. After an initial sift, 31 teams prepared formal applications.

There needed to be a way to assess applications consistently, and in line with the guiding principles of the UTC movement such as university and employer sponsorship. DfE officials and their BDT colleagues reviewed the completed forms before panel interviews were held with applicants. Charles Parker, at that time BDT's director of operations, explained:

> We persuaded officials that the whole process should have a high degree of Baker Dearing involvement. I remember being completely explicit about the fact that we had an overlapping interest. At first, I assumed that the Baker Dearing people who'd been promoting the project and helping the applicants would not sit on the other side [with DfE officials] when applications were reviewed. But we were asked to work out a pecking order within a batch of applications, from best to worst,

scoring them as rigorously as we could. The Department would do the same, and then we would go to a meeting with our respective pecking orders. We were seldom far apart. The only disagreement was really where the line should be drawn, and usually we were only talking about one or two either side of the line.
(Parker, 2018)

After all the interviews had been completed, recommendations were sent to the Secretary of State, Michael Gove, for his consideration. Initially, 13 were approved and included in a list announced by the Secretary of State, Michael Gove, on 10 October.

- Bristol and South Gloucestershire: specialisms, engineering and environmental technology
- Aylesbury: construction and IT
- Burnley: engineering and construction, and building technologies and the environment
- Houghton Regis (Central Bedfordshire): engineering and manufacturing
- Daventry: environmental sustainability, sustainable construction and new technologies in engineering
- Liverpool: life sciences, including healthcare diagnostics and medical equipment supply
- Newcastle: science, technology, engineering and maths
- Nottingham: science, technology, engineering and maths
- Plymouth: marine engineering and advanced manufacturing
- Sheffield: advanced engineering and creative and digital industries
- Silverstone: high performance engineering and motorsport, and event management and hospitality
- Southwark: medical engineering and health technologies, and construction and property management
- Wigan: manufacturing engineering (food production) and green energy and environmental technologies

A fourteenth UTC, UTC Reading, was added to the list shortly afterwards.

In his statement to the House of Commons, Michael Gove said:

[T]here is a new model of academy whose development has the potential to be particularly transformational – the university technical college. Thanks to the leadership shown by Lords Adonis and Baker, and the vision of Sir Anthony Bamford of JCB, the first university technical college opened its doors in September last year. Educating young people from the age of 14 to 19, with a curriculum oriented towards practical and technical skills, with support from industry and sponsorship from a university, these schools have the potential to transform vocational education in this country immeasurably for the better. They combine a dedication to academic rigour – with the JCB UTC delivering GCSEs in English, maths, the sciences and modern languages – with the adult disciplines of the workplace. Longer school days and longer school terms contribute to a culture of hard work and high aspirations.

The JCB UTC was joined by another [Black Country] in Walsall this September, and three more are in the pipeline. If we are to ensure that the benefits of UTCs, academies and free schools reach many more children we have to up the pace of reform. That is why I am delighted to be able to announce today that my Department has given the go-ahead to 13 new UTCs in Bristol, Buckinghamshire, Burnley, Bedfordshire, Daventry, Liverpool, Newcastle, Nottingham, Plymouth, Sheffield, Southwark, Wigan and at Silverstone race track. This Baker's dozen of UTCs will specialise in skills from engineering to life sciences, and I am convinced they have the potential to change the lives of thousands for the better.
(HC Deb 10 October 2011, col. 63)

The announcement was supported on all sides of the House. The Shadow Secretary of State for Education, Stephen Twigg, described UTCs as an exciting innovation, while warning that they risked being undermined by the requirements of the English Baccalaureate. Graham Stuart, Conservative backbencher and chair of the House of Commons Education Committee, welcomed support for more

UTCs and urged that for those young people not living near a UTC, transfer to a further education college should be an option. Frank Field (Labour) said the country needed not 13 additional UTCs, but 113. Robert Halfon (Conservative) described UTCs as 'an essential instrument of social justice' (HC Deb 10 October 2011, col. 67).

BDT held an overnight seminar for the successful applicants soon afterwards. Officials from the DfE, Young People's Learning Agency (YPLA) and Partnerships for Schools (PfS) attended to ensure every newly-approved UTC understood the process that must be followed prior to opening. (The functions of the YPLA and PfS were transferred to a new body, the Education Funding Agency, in April 2012.)

Key steps included:

- Establish an academy trust to run the UTC: the DfE prepared a model memorandum and articles of association for partners to use
- Compile a detailed education brief, including a curriculum plan, admissions policy, space requirements and an accommodation schedule
- Conduct a public consultation in the proposed UTC's catchment area, including the specific question as to whether the Secretary of State and proposed academy trust should enter into a funding agreement and the UTC should open
- Prepare a project plan with PfS, the agency responsible for overseeing the procurement of land, buildings and infrastructure (including information technology) for new schools: each UTC project was allocated a PfS project director and an ICT advisor
- Secure planning approval
- Submit a final business case to ministers, leading to a firm commitment to capital funding
- Sign a funding agreement with DfE to cover the UTC's running costs.

A DfE official acted as project lead for each new UTC. A team of education advisors provided support to UTC project teams and principals and reported progress, issues and concerns to the project leads. Their reports were discussed with BDT colleagues at fortnightly meetings.

In November, Lord Baker hosted a reception for representatives of all UTCs then open or approved. The Minister of State for Skills and Lifelong Learning, John Hayes, spoke to the guests:

> We warmly endorse the work which is being done on UTCs, led by Ken Baker. Without his commitment, drive, enthusiasm and expertise, the programme would not have been the success it has already been and will continue to be.
>
> There have been three great Secretaries of State [for Education]. One was Rab Butler, whose 1944 Act paved the way for technical schools. The next is right here, Ken Baker, who is giving life to Butler's original vision as he did when Secretary of State – many of you will remember his role in establishing the first City Technical Colleges. Now we have a Secretary of State [Michael Gove] who has taken this on, too. Three Secretaries of State committed to creating high-quality secondary education through schools that specialise in practical and technical learning.
>
> My vision is to elevate the practical – to blow apart the misapprehension that the only way of gaining prowess is through academic learning. All of those with technical tastes and talents, with practical aptitudes, deserve their place in the sun, their share of the glittering prizes. That is about creating a sustainable economy, of course, because we need those high-tech skills. But it is also about creating a society which is sustainable, in which each feels valued so all feel valued; a society in which people start the day with a spring in their step because their work involves prowess and a sense of self-worth.
>
> (Hayes, 2011)

Ministers were so enthusiastic about UTCs that they agreed to open more than the 24 previously pledged by George Osborne. In 2012, a further 15 applications were approved:

- Swindon UTC: engineering with business, enterprise and entrepreneurship

- UTC Bluewater (Kent): engineering and integrated computer science
- Heathrow Aviation UTC: aviation engineering
- East London UTC (Elutec): mechanical (manufacturing) and engineering and product design
- Norfolk UTC: advanced engineering and energy skills
- Elstree UTC: entertainment technologies and crafts, electronic engineering and digital technologies
- Harlow UTC: environmental engineering and medical technology
- UTC Cambridge: bio-medical science and technology, and environmental science and technology
- Lincoln UTC: engineering and core science
- Warwick UTC: engineering (with digital technology)
- West Midlands Construction UTC (Walsall): construction and the application of information technology in the built environment
- Birkenhead UTC engineering
- Liverpool Low Carbon and SuperPort UTC: engineering and logistics
- Energy Coast UTC (Cumbria): energy
- MediaCityUK UTC (Salford): creative industries and entrepreneurship

9
RECOGNIZING SUCCESS

The Duke of York Award for Technical Education was developed as a way of recognizing the breadth of UTC students' achievements through an award of their own. An outline was developed by BDT and refined in consultation with Jim Wade, principal of JCB Academy. It took account of previous awards, including in particular the Kingshurst Achievement Award, developed by the first City Technology College. Kingshurst's former principal, Valerie Bragg, explained the award in these terms:

> The Kingshurst Achievement Award was offered as a Certificate for the under 16s and Diploma for the over 16s and awarded at three levels – gold, silver and bronze. We commissioned a survey, which was carried out for us by Birmingham City University, to find out what skills employers wanted of their recruits. The result became our 'core skills'. We then asked the University to work with us to develop the Award.
>
> We produced an Award booklet for each student, containing lists of things that students could do, including the mix of hard and soft skills that employers said they wanted. The booklet was validated by major companies [that worked with us] and also by the University. Booklets had to be signed off over the student's years at the CTC. The actual certificates were signed by the University and carried companies' logos. We developed

'compacts' with the University, which gave places to students who had achieved the Diploma, and with employers – who very much wanted our students.

The Kingshurst Achievement Award was highly regarded. If a similar thing were produced for the UTC movement, I think it would be even more desirable, powerful and beneficial.

(Bragg, 2012)

The proposed 'UTC Award' would be administered locally by individual UTCs and co-ordinated nationally by BDT. Students would be eligible for an award at one of four levels: certificate, bronze, silver and gold. Each level would recognise a combination of qualifications, experience and skills. An important element of the scheme would be evidence provided (or endorsed) by people outside the UTC, such as local employers and students and staff from the sponsor university.

By way of example, the requirements for a bronze UTC Award might include:

- an approved level 2 technical or vocational qualification which required at least 120 guided learning hours – that is, the same size as a single GCSE
- grade A* to C passes in at least four GCSEs including English and maths
- evidence from someone from outside the UTC confirming that the candidate–
    - had kept a record of at least four team projects
    - was able to describe (in conversation) what they had learned as a result of taking part in team projects, and how they had come up with and tested ideas for solving a problem during a team project
    - had completed a work experience placement of at least one week
    - had clear plans for what they would do after the age of 16 and could describe the options and careers open to them in future.

Ideas for the UTC Award were being formed at a very interesting moment. For one thing, HRH The Duke of York had started to take a close interest in UTCs; and for another, the Gatsby Charitable Foundation had asked Sir Mike Tomlinson to help connect the UTC curriculum with ladders of professional recognition in science, technology, engineering and maths (STEM).

HRH The Duke of York visited Hackney Community College in October 2011 as part of his efforts to promote apprenticeships, particularly for school leavers. He toured the Shoreditch campus, was briefed on the College's Tech City Apprentice Programme, and heard about plans for a UTC that was due to open in autumn 2012. He immediately agreed to host a dinner to brief potential business partners about the UTC and local apprentice initiatives.

The Duke wanted to hear more about the UTC movement and in February 2012, hosted a roundtable discussion with representatives of 19 open or planned UTCs, together with Lord Baker and Charles Parker. Plans were made for him to visit other UTCs in the months to come.

BDT also considered asking the Duke to support an outstanding achievement award for UTC students, linked to the proposed UTC Award. Consultation with UTCs showed strong support for the idea; more than that, it was suggested that the whole award scheme might bear the Duke's name. BDT's trustees supported the move, and after a meeting at Buckingham Palace, the Duke gave his approval in principle for a Duke of York Award for Technical Education.

Sir Mike Tomlinson and Nigel Thomas (Gatsby Charitable Foundation) saw a possible link between the proposed Duke of York Award and work they had been carrying out with the Engineering and Science Councils. Lord Sainsbury wrote to Lord Baker about this in July 2012:

> Over the last six months Gatsby has been supporting Mike Tomlinson to work with the Engineering Council, Science Council and key professional bodies to develop a 'pre-registration' framework within UTCs.
>
> Such a model could provide professional recognition, perhaps at two levels (one aimed at 16-year olds, the other at 18-year olds) of technical knowledge and skills alongside important transferable skills sought by employers ...
>
> I understand ... that HRH The Duke of York is interested in sponsoring certificates of achievement

within the UTCs at three levels. If managed carefully, the gold and silver level Duke of York certificates could be the awards for the two pre-registration grades being developed. It would clearly be folly to have two separate schemes vying for support within the fledgling UTC system and a single scheme, with both royal endorsement and the imprimatur of the professional engineering and science bodies, could be particularly aspirational.

(Sainsbury, 2012)

Lord Sainsbury wanted UTC students to appreciate that they were on a path towards professional recognition as technicians, scientists and engineers.

Many UTC alumni were likely to become Chartered Engineers (CEng), Incorporated Engineers (IEng) or Registered Engineering Technicians (EngTech) at some point after leaving their UTCs. Others would become members of professional bodies in the fields of science and ICT.

Historically, gateway qualifications for CEng included Higher National Certificates and Diplomas. After 1970, entry was restricted to engineers with honours degrees, and since 1997 all CEng candidates have had to demonstrate a combination of higher learning (a master's degree, for example), experience in the workplace and continuing professional development.

After 1970, people with higher nationals or equivalent engineering qualifications could apply for recognition as 'Engineering Technicians' or (confusingly) 'Technician Engineers'. During 1987, these terms were phased out in favour of Incorporated Engineer (IEng). That still left a gap for technicians with qualifications below the level of higher nationals, and in 2003 a new scheme, EngTech, was introduced for engineering technicians with skills and qualifications at level 3 gained either through an Advanced Apprenticeship or a course leading to a recognized level 3 engineering qualification and a relevant job. Having achieved EngTech status, an individual might aspire to IEng or even CEng status at some point in the future.

A common competence framework was developed for EngTech, IEng and CEng, covering –

- Knowledge and understanding
- Design and development of processes, systems, services and products
- Responsibility, management or leadership
- Communication and inter-personal skills
- Professional commitment

Importantly, recognition was not based solely on candidates' qualifications, but also took account of their wider skills and experience gained in the workplace.

Work on the EngTech framework was overseen by the Engineering Council, which brought together all professional engineering institutions operating in England. The Council went on to develop a similar framework for ICT technicians, called ICT-Tech, in 2009.

Lord Sainsbury took a close interest in this ladder of opportunity, not least because of concerns about impending shortages of technicians in the STEM sectors. In 2010, he presented a report on the professional registration of technicians to the then Secretary of State for Business, Innovation and Skills, Lord Mandelson (Sainsbury 2010). He concluded that:

> Creating a common framework of registration for technical staff working across the science, engineering, health and ICT sectors presents a huge opportunity to establish and maintain common quality standards for UK technicians and ensure that the skills and knowledge learnt within technician pathways are in line with employers' needs.
> (Sainsbury, 2010, p. 9)

Among other things, Sainsbury recommended creating a register for science technicians (RSciTech) and establishing a Technician Council to promote all three registration schemes – EngTech, ICT-Tech and RSciTech. His recommendations were accepted and the government provided seed-corn funding for the new Technician Council.

Soon after he presented his report to Lord Mandelson, Sainsbury spoke at a conference, *Technical Education for the 21st Century*, which was co-hosted by Gatsby and the Edge Foundation. He said:

It is predicted that by 2020 we will fall short of our target for the number of people in the workforce with Level 3 qualifications by some 3.4 million. Of these, a significant proportion will be in the STEM-related sectors. These people, the technicians of the 21st century, will be key to the decommissioning of our ageing nuclear power stations and the construction and maintenance of new ones; to ensuring that the switchover to digital television occurs on schedule; and to manufacturing the high-tech products we will need to sell to the rest of the world in the future.
(Sainsbury, 2011, p. 1)

The Gatsby Charitable Foundation became a major supporter of the Baker Dearing Educational Trust, making grants to support BDT's operational costs and nominating Nigel Thomas to serve as a BDT trustee.

Sir Mike Tomlinson's working group developed a set of standards to be used to certify the competencies and experience of students in the 14-18 age range, linked to the more advanced skills and experiences required for full EngTech, ICT-Tech or RSciTech registration. These related directly to the standards for registered technicians, including:

- Knowledge, understanding, application, action and professional practice
- Accepting and exercising personal responsibility
- Communication and interpersonal skills
- Professional standards

The working group devised standards at intermediate level and advanced level – broadly equivalent to levels 2 and 3 – and suggested piloting them at selected UTCs. In time, they would be available to all UTCs and to other schools and colleges with specialist programmes in science, engineering and technology.

BDT worked with Sir Mike Tomlinson, Gatsby and professional bodies to incorporate the pre-registration framework into the proposed Duke of York Awards. It was, in fact, a very good fit, and provided a way to describe and assess the skills and abilities developed through the UTC curriculum – particularly the cross-curricular projects developed and delivered in partnership with employers and sponsor universities.

BDT's curriculum working group agreed that the Duke of York Award – now incorporating a lightly amended version of the pre-registration competencies framework – should be piloted by the JCB Academy and Black Country UTC. HRH The Duke of York granted his approval, and the Duke of York Award for Technical Education was announced on the day of his visit to Black Country UTC, 8 July 2013.

Lord Baker said BDT was honoured that The Duke of York had lent support to the new Award, which would:

> ... only ever be made to students who follow a rigorous technical curriculum, supported by real-world projects, challenges and work experience with leading employers. The Award will encourage holders to aspire to become professionally registered engineers, technicians or scientists in the future.
> (Baker Dearing Educational Trust, 2013a)

He also announced that The Duke of York had agreed to be patron of the Baker Dearing Educational Trust, a position he held until the end of 2019.

The Duke of York Award was piloted by the JCB Academy and Black Country UTC in 2013. It was not compulsory for students to put their names forward for the award. There was an element of guidance from staff, too – particularly at Black Country UTC, where 37 year 11 students were not put forward for awards or chose not to be interviewed. Similarly, a number of year 13 students were not put forward or chose not to attend scheduled interviews.

In total, 162 candidates were interviewed by one of six members of the BDT team. Questions explored elements of the competency framework originally developed by the working group chaired by Sir Mike Tomlinson.

A report was prepared for BDT's curriculum working group:

> At the outset, the competency framework was the greatest unknown quantity. While well-written and well-constructed, we could not be sure if it was

entirely relevant to UTC students. The interviews showed it to be a very good match to the wider skills and abilities fostered by the UTC curriculum such as teamwork, problem solving and finding, analysing and applying information.

(Harbourne, 2013)

Among year 11 students (age 16), standards were high at both UTCs, with most students achieving the silver standard. The only general weakness concerned knowledge of professional institutions. Most students recalled learning about professional bodies, trade associations and unions during year 10, as part of their Principal Learning Qualification. However, memories were often hazy: some students mentioned trade unions first and only mentioned relevant professional bodies when prompted. Knowledge also varied among year 13 students, though most of those who planned careers in engineering said they expected to join a professional institution in the future and some were aware of different levels of membership.

Interviewers were impressed not just by candidates' knowledge and experience, but also by their confidence and capacity for reflection. Year 11 and 13 students all spoke with pride about the projects they had worked on, explaining clearly what had worked well and how they had responded to problems and challenges. Students were asked about their experience of working in teams, and almost without exception they found they had worked well together: only occasionally had team members not pulled their weight or seriously fallen out with one another.

Some students had learning difficulties and/or disabilities, the most common being dyslexia. Interviews for the Duke of York Award took no account of reading and writing skills, and dyslexic students performed well – entirely as expected. A student with more profound difficulties also impressed at interview, showing pride in what he had achieved and a clear understanding of his plans for the future. Staff later told the interviewer that this student rarely spoke when he started at the UTC, but was now much more confident and happy to talk about his progress. He expected to work in his father's business after leaving the UTC. Summary notes were prepared after each interview, including these examples:

Notes from an interview with a year 11 student at Black Country UTC: [Candidate A] acted as team leader for the ZF Lemförder project, working on ball joints. [ZF Lemförder manufactures components for the automotive industry.] She set the agenda for team meetings and allocated tasks relating to planning, health and safety, materials, manufacturing and quality control. There were disagreements within the team about roles and responsibilities, but [candidate] talked things over with team members and resolved differences.

The team learned to use a CNC machine to manufacture components to required tolerance. They made four, examined and critiqued them, and taking account of results so far made one final component which met specifications. She was pleased with the final outcome and lessons learned.

[Candidate] spoke knowledgeably about codes of conduct including health and safety, personal protective equipment and risk assessments. She was aware of professional bodies, citing Institution of Electrical Engineering in particular.

[Candidate] came to UTC because of its science specialism but is now more interested in engineering. She will remain at Black Country for post-16 studies. Long term, she wants to specialise in applied medical engineering.

Notes from an interview with a year 13 student at JCB Academy: [Candidate B] led on two projects. Team met every day. Spoke about the difference between 'leader' and 'boss': as leader, he had responsibility for ensuring project was completed on time and to specification, but did not stand over team members to supervise their work – encouraged them to take responsibility for own actions.

[Candidate] described applying maths and physics to projects: e.g. expansion/contraction of materials in engine where exhaust fumes reach temperatures of up to 550°C; turbine production costs also mentioned. Used a lathe and bench skills to make a hole punch.

Worked under pressure: e.g. a presentation to Bentley Motors had to be re-scripted when a team member failed to arrive on time. Carried out market research into reasons people choose one mobile phone over another, and wrote up findings in a 5000 word report. Attended workshop on disaster management and recovery run by university. Described steps to become a Chartered Environmentalist, which is what he wants to be – degree, master's, four years' experience, perhaps a PhD – very interested in why and how volcanoes develop and change shape.
(Baker Dearing Educational Trust, 2013b)

Students were asked about their plans for the future. Among year 11 students, the most common responses were (a) stay at JCB/Black Country for post-16 studies or (b) start an apprenticeship, usually in engineering, or (c) change direction entirely, which usually necessitated moving to other schools or colleges to pursue subjects other than engineering. For example, one student planned to train as a chef; another wanted to take A-levels with a view to becoming an English teacher. Year 13 students mostly planned to go to university or start a higher apprenticeship. Many were spoiled for choice, having received multiple offers, and some declined places at university in favour of apprenticeships with leading businesses.

On 9 December 2013, 160 students from the JCB Academy and Black Country UTC travelled to Buckingham Palace to receive awards from HRH The Duke of York. In addition to bronze, silver and gold awards, the Duke presented two outstanding achievement awards.

The first was presented to Joe Symonds. Joe could have stayed at his previous school to study for A-levels, but instead transferred to Black Country UTC to study engineering, physics and maths. Having started at the UTC with no knowledge of robotic control systems, Joe made such rapid progress that he gained a place on the UTC's mechatronics team and took part in the national finals of a competition organized by WorldSkills UK. Most members of competing teams were older than Joe and were either studying at university or had jobs in mechatronics.

In his time at Black Country UTC, Joe was also managing director of a Young Enterprise team that won a number of awards and represented the UK in Sweden. In addition, he participated in a mentoring programme for Year 11 students and provided one-to-one tuition in mathematics for a number of pupils.

At the end of his time at Black Country UTC, Joe secured a Higher Apprenticeship at Jaguar Land Rover, a six year programme combining work with study at college and university, leading to a degree and – just as importantly – the knowledge, skills and experience needed to be a first-rate engineer.

The second award was made to a team of students from JCB Academy. During the Toyota Challenge, key stage 4 students were given the task of fitting fuel spacers to an engine in a set amount of time. Students had two weeks to develop their ideas before competing under timed conditions, in front of judges from Toyota UK.

The winning team – Georgia Turner, Callum Edge and Tom Bather – used Kaizen principles, which encourage people involved in a process to come up with ways to improve it. Their solution proved to be entirely new, and involved making a jig to fit the spacers in quick succession. Come the day of the timed trial, they fitted all 16 in just under three and a half minutes – much faster than any other team.

The Deputy Managing Director of Toyota UK was so impressed that he asked his staff to investigate the students' solution further and did not rule out incorporating their ideas into Toyota's own manufacturing process.

Speaking to *The Independent*, HRH The Duke of York said:

> Often, employability skills in their widest range have not been included in what people have learned. What we've got to do is try to give young people not only that fundamental education, but the ability to be able to go into the workplace with greater confidence than they would previously contemplate.
> (Garner, 2013)

THE DUKE OF YORK STEPS DOWN

The Duke of York stepped down as patron of the Baker Dearing Educational Trust in 2019. BDT's board of trustees gave the award scheme a new name – the Baker Award. BDT's chief executive at that time, Simon Connell, said:

I think [the Award] is absolutely fantastic. We offer it to UTCs free of charge. It requires relatively little extra work from UTCs, it recognizes 'UTC-ness', and it's a good indicator to us that UTCs have an effective curriculum.

Gold Award students go to the ceremony at an important time – after they've left their UTC, when their memories of the UTC are still powerful. It's an opportunity to catch up with their friends and it's a really uplifting experience for all of them. They talk honestly and fondly about their time at the UTC.

The 2020 ceremony will be broadened into a wider celebration of student achievement, to include UTC students who have done other amazing things. It's an opportunity to celebrate everything that's great about UTCs.

(Connell, 2020)

10

LIVERPOOL LIFE SCIENCES UTC

Liverpool Life Sciences UTC was the brainchild of former education publisher, Nigel Ward. He told the author how the idea came about.

I was involved in educational publishing for a long time – cognitive abilities tests, curriculum software and so on – and then I headed Collins Education. I was close to schools for 20 years and then in 2003 I was approached by a number of people which led to a meeting with Andrew Adonis, who asked me to work with Stanley Fink [chief executive of Man Group plc] and get involved in the sponsored academies programme. I came to Liverpool and visited two failing schools and a few months later, we set up the North Liverpool Academy Trust, which later became the Northern Schools Trust.

The North Liverpool Academy opened in 2006, replacing two schools which had been failing for generations. We moved to a fabulous new building – which I project managed – in September 2009. I took an 18 month sabbatical to supervise the new build. The school has done very well. Since 2006, the outcomes have been good – on a par with national outcomes – providing quite a boost to the community. It's educationally and pastorally very strong. It's an aspirational school and in Ofsted terms, good with outstanding features.

Since taking on that school, I've always wanted to push the envelope and provide a curriculum which engages all young people, because that's the way to get the best out of them, and also how you improve attendance. We reached out to local universities and businesses and probably as early as 2006/7 we started to provide enrichment activities with some local software developers who were making high-end games for Sony and Nintendo. We realized we had some students in the school who were attending a traditional state school but earning money in their spare time writing software for international games companies. Once that became apparent, we decided to beef up the enrichment and that led to us having our own school within a school, a programming technology school.

I then wanted to make that a much bigger proposition. At the time, the whole 14 to 19 debate was raging. There's a lot in our sector about 'farming' young people to meet particular targets and objectives, almost showing scant regard for the next stage. Well, in my schools, we take a much wider vista. We nurture them, keep them safe and prepare them for the next stage. About 45 per cent of our students go on to do vocational programs when they leave.

Nevertheless, I felt we were taking them only so far because our offer was a bit lacking. The question was, what are we going to do about that next stage? Local provision was a mixed bag and some providers were in special measures.

So all these things were colliding: me wanting more vocational routes, the academy gaining experience through the programming school, and then becoming aware of the work Kenneth Baker was doing. I knew of his work with CTCs because they used to be my customers: I used to do their testing for them. I heard of his plans for UTCs.

We decided to do some research, based on thinking about paths to employment. What kind of schools would we need in Liverpool to prepare young people for good, local jobs? That's the businessman's approach, as opposed to the typical educator's response – there's a chance to open a school, so let's open it. We spent three months on quantitative and qualitative research and came up with two industry sectors that gave us everything we needed: plentiful jobs, good wages, a good career and career mobility. There were big enough companies that if our students got into them, there would be mobility within and around the industry. Those two industries were first, technology and gaming and second, life sciences and bioscience. Sony is a big presence locally and within 25 miles of Liverpool is probably the largest concentration of biotech research companies and pharma companies in the UK. We decided the UTC would specialize in life sciences and that we'd open a Studio School to focus on gaming and programming. And then we bid!

The two bids were submitted within four weeks of each other because we expected to get one but not both approved. That way, we would make one of them work and come back to the other later. In the event, we got approval for both.

Corralling employers was something I found easy, with my private sector background. I knew what they wanted, and what they didn't want. I was able to pitch the idea very clearly. We showed businesses that we understood their challenges and said, 'these are the problems I can solve for you'. That's different from saying 'this is what the school is going to be'. It's about selling the benefits, not the features.

I went to see some senior people in a global drugs company. I told them I could slash their graduate recruitment costs. They said, 'we're listening – you've got half an hour'. We won them over.

We also benefited from other people's networks. A couple of governors in biotech introduced us to their friends and friends of friends, for example. We then hosted a series of meetings and dinners – some in the office, some in the pub – where we talked to all the biotech businesses about what we planned to do. We started to get a groundswell of support. Not only that, but it fed back into academia because many of the bioscience businesses have themselves spun out of academia to a certain extent.

Combined with that, North Liverpool Academy was in a partnership with Liverpool University. We have a wonderful, really deep relationship with Liverpool University. We also work with John Moores University, UCLAN [the University of Central Lancashire], Dundee University, Bradford University and Sheffield University. You name it, we work with them.

Our links with Liverpool University translated into masterclasses, input into the curriculum and PhD students who train as teachers to work in the UTC. Bioscientists who sit on our board of governors [today] are now asking us to help them with a challenge: preparing BTEC students to survive and thrive as science undergraduates. We've been exploring this with Liverpool and Sheffield universities independently, looking not just at how we teach the BTEC, but which modules we teach depending on the anticipated destinations of the students. In other words, we've gone from a simple supply side relationship, where we were looking for help from the universities, to a more balanced relationship where we help each other.

(Ward, 2017)

EDGE CASE STUDY

The Edge Foundation published the following case study about Liverpool Life Sciences UTC in 2014, a year after it opened.

Liverpool Life Sciences University Technical College (UTC) is the first science-based UTC in the country and is already a phenomenal success. With room for 200 students in each year group, applications have flooded in from students across Liverpool and beyond – 90 schools in total – and some students travel up to an hour each way.

The UTC is housed in the Contemporary Urban Centre, a listed Victorian warehouse which was converted first into an arts centre and then into the home of two innovative schools. The Studio is a 14-19 studio school specializing in creative media, gaming and digital technology, while Liverpool Life Sciences UTC is the first school in the UK specializing in Science and Health Care for 14 to 19 year olds. Both schools are part of the Northern Schools Trust.

The UTC curriculum ensures students combine core subjects with their chosen technical specialism. On average, two hours of each day is dedicated to project-based learning.

Here in Liverpool, students work in project teams on employer-inspired projects that combine science, healthcare, maths and English. The UTC has industry-standard laboratories and equipment – in some cases, better than the facilities available to university undergraduates.

Most schools, colleges and even universities deliver experimental science through set-piece laboratory classes. Liverpool Life Sciences UTC has taken a different approach, inspired by the Director of Research and Innovation, Professor David Hornby. He says, 'At university, students usually carry out their first research projects in the lab after their second year of study. I decided to ask UTC students to design their own projects, and the ideas that came back were exceptional.

'For example, we are working with the University of Liverpool

Genome Centre to sequence the mealworm genome, supported by a grant from the Royal Society. It's the first time a school has been involved in a project as advanced as this.'

He added: 'Students will fondly recall the first tub of 99p dried mealworms and the 6 pack of [a drink called] Poweraid: my best source of a negatively charged brilliant blue dye in chromatography classes!' After that, year 12 students set up their own meal worm farm to provide a permanent supply of their own.

The project has taken off in some unexpected directions. For example, students realized they needed to create, store and analyse large numbers of samples, which meant buying expensive racks and other industry-standard laboratory equipment.

Using 3D printing, the Greenland Biodesign team – a group of post-16 students – designed and fabricated a range of customized sample holders that accommodate tubes ranging from the size of a coffee bean to tubes as big as a banana. Soon they were printing lab equipment such as column chromatography racks and a fully operational, battery operated microcentrifuge – and all for a fraction of the normal price.

Elsewhere, healthcare students have access to fully functioning hospital beds, complete with 'Sim Man', a dummy patient who displays vital life signs and symptoms, and all the equipment needed for checking the status of patients. There's also a mock nursery and home setting for practical studies of health, social care and childcare.

The UTC and Studio School are co-located and share some facilities, including a 120-seat cinema/theatre. Year 10 students went there to watch a Royal Shakespeare Company production of Henry IV Part 1: they enjoyed it so much that now they can't wait to see Part 2! The cinema is also used for masterclasses by leading speakers from around the world. The UTC principal, Phil Lloyd, describes the masterclasses as his version of TED Talks.

The UTC has tremendous support from leading employers including Unilever, Novartis and Croda. All post-16 students also have the opportunity to complete work placements with another sponsor, the Royal Liverpool and Broadgreen University NHS Trust.

Students record their progress and achievements in a skills passport, including transferable skills such as project planning, data analysis, computational skills, and numeracy and literacy.

Looking back at the first year of Liverpool Life Sciences UTC,

Prof Hornby sums up his experience in one simple sentence: 'Never, ever underestimate the ability of young people.'
(Harbourne, 2014)

LIVERPOOL LIFE SCIENCES UTC IN 2020

In 2020, the Northern Schools Trust wrote:

> [Liverpool Life Sciences UTC] provides outstanding academic and vocational education by working closely with local employers to create the next generation of scientists, healthcare practitioners, engineers and entrepreneurs.
>
> Students have access to world class resources and facilities, and a curriculum developed in conjunction with business partners and delivered by experts both in and out of the classroom.
>
> Life Sciences UTC is the only state school in the country to employ University Professors and PhD students as part of the teaching faculty.
>
> Partners include University of Liverpool, Siemens, Novartis, Unilever, Pro Lab Diagnostics, 2Bio and Liverpool Community Health to name but a few.
> (Northern Schools Trust, 2020)

Ofsted inspected Liverpool Life Sciences UTC in January 2020, when there were 476 students on roll – 216 in key stage 4 and 260 in years 12 and 13. The principal, Jill Davies, is head of both the UTC and the Studio School (The Studio, Liverpool) and is supported by a single senior leadership team. Some core subjects are taught jointly across the two schools.

Ofsted's inspectors concluded that the UTC was 'good', and praised careers advice and sixth form programmes especially highly:

> The principal and governors are highly ambitious for all pupils to do the best that they can. The school provides an exciting curriculum with a range of

academic and vocational subjects, including English and mathematics.

Leaders place a strong emphasis on building pupils' research and practical skills.

Careers advice is highly effective. There is a strong emphasis on developing pupils' employability skills, such as teamwork and communication. Pupils and students benefit from mentoring and presentations from local employers, hospitals and universities. This helps to raise their aspirations for their future careers.

Students in the sixth form achieve exceptionally well in vocational courses. Leaders have improved the curriculum planning and delivery in the academic subjects. As a result, students are performing better than in the past and achieve well. Students value the support that teachers provide with applications to further education and apprenticeships. Many go on to study at university.

(Ofsted 2020a, pp. 1-2)

Nigel Ward was especially proud of students who joined the UTC as – in his words – 'reluctant students':

Engineering UTCs stimulate reluctant students with work which is tactile and physical. Our life science laboratories have the same effect: for disaffected boys from Toxteth, practical work is transformational. The fish in our fish tanks enrich the science curriculum, but just looking after them draws the students in – almost covertly. The same goes for the plants we're growing in the basement, using hydroponics. You've got to find the key to the door.

(Ward, 2017)

THE STUDIO, LIVERPOOL

Like UTCs, Studio Schools were set up to offer specialist 14-18 programmes of education. They particularly emphasized entrepreneurial skills. Many specialized in aspects of the digital economy, including design and IT. The prospectus issued by The Studio, Liverpool, stressed the development and application of skills:

> The Studio, Liverpool educates young people for success in a digital world – in particular for employment, entrepreneurship, a new business venture or further study in the digital media sector.
>
> Our students are creators, not consumers. Technology is the enabler – it helps us create and achieve things, rather than being the end goal in itself. This empowers them to generate opportunities for themselves within this emerging, exciting sector.
>
> Some students apply their skills directly in the gaming, animation, digital marketing and computer technology sectors. Some are artists, creatives and musicians, while others will work for businesses and organisations that use digital media to develop products and services and connect with clients, consumers, patients and citizens.
>
> They are the future digital leaders, using their expertise to improve the lives of others through digital technology.
>
> The Studio feels and works in a different way to a conventional school. We have a longer school day, that more closely resembles full-time work, and closer relationships with this sector and the community. You'll develop expertise and broaden your horizons by working with the innovative businesses that are part of that community. And you'll be given responsibility for your learning and progress, alongside the support you need to make the very best of these opportunities. At The Studio, you won't learn how to react to a changing world; you'll create a changing world, for the better.
>
> (The Studio, Liverpool, 2018, p.6)

11

UTC READING

Rob Wilson was elected to Parliament as the Conservative member for Reading East at the 2005 general election. He said:

> In my view, secondary education in certain parts of Reading – and in particular, my own constituency – was not as successful as it should be at that time. Until 2010, there was very little I could do about it, because I wasn't part of the governing party. After that, the coalition was formed and Michael Gove was appointed Secretary of State for education. He was committed to opening free schools and giving parents control. I thought I could be the catalyst to help bring about changes in Reading that parents wanted and needed. It was just a question of how to do it.
> (Wilson R., 2018)

As part of his efforts to improve educational outcomes, Wilson formed a committee to push for a new school in Reading East. Members included representatives of local businesses, ranging from very large companies – Network Rail, Microsoft and Cisco, for example – to locally-based firms such as Peter Brett Associates, a firm of consultants working on major development and infrastructure projects. They were enthusiastic about establishing an 11-16 secondary school with a technical curriculum. However, the Department for Education made it clear that any new school should also have a sixth form.

Lord Baker heard about the plans and arranged to meet Rob Wilson. He explained the UTC concept and encouraged Wilson to put forward an expression of interest (EoI).

The first EoI was submitted in May 2011. It was drafted by Lee Nicholls, Vice Principal of Oxford and Cherwell Valley College. The college was part of a group of colleges, including Reading College, which later adopted the umbrella name 'Activate Learning'. The college was named as lead sponsor, with Reading University as co-sponsor. It was hoped that Reading and Wokingham Borough Councils would also be co-sponsors. Employer supporters would provide technical advice and support.

The EoI set out plans for both a 14-19 UTC and an 11-14 academy on the same site, a former secondary school which had more recently been used to deliver further and higher education. The academy would offer a broad-based educational foundation at key stage 3, with an emphasis on applying learning to real life, while the UTC would specialise in computing and construction engineering.

The EoI explained the need for additional key stage 3 places in East Reading. At that time, the majority of local children attended secondary schools just across the border in Wokingham, but demographic growth was already leading to a shortage of places. Having a new 11-14 academy in the area would help fill the gap, providing a new option within easy walking distance for many local families. At the end of key stage 3, pupils would have the option to move to other secondary schools or to the linked UTC. UTC recruitment at 14 and 16 would extend over a wide catchment area covering Wokingham, Reading and West Berkshire.

The proposals were turned down because of the policy focus on establishing 14-19 UTCs. In Lord Baker's view, young people were better able to choose a technical curriculum at 14 than at the age of 11. The EoI also fell short of the expectation that employers would act as UTC sponsors and would appoint a significant number of governors. Rob Wilson added:

> I wasn't certain whether we could get Reading and Wokingham [Councils] to support the idea that young people would switch schools at 14. That was my big

concern. Parents weren't used to it and there was no tradition of it in recent history.

(Wilson R., 2018)

At about that time, Labour took control of Reading Borough Council and took the position that any new school should be 11-16, not 14-19. With support from Wokingham Borough Council, however, there was sufficient impetus to carry the project forward.

A revised application was prepared, giving employers a greater involvement in the governance of a 14-19 UTC. By the end of 2011, the proposal seemed certain to be approved.

Sir David Bell was appointed Vice Chancellor of the University of Reading on 1 January 2012. He had previously been Permanent Secretary at the Department for Education. Taking stock, he felt it would be better for the university to support many local schools than to sponsor individual institutions such as the UTC. News that the university was no longer prepared to act as co-sponsor came as a blow, as it was normally a prerequisite for all new UTCs to be sponsored jointly by a university and local employers.

BDT immediately took soundings to see if another university could be persuaded to step in at the 11th hour. By the beginning of March, however, a compromise was agreed, naming Reading University as a *partner*, rather than *sponsor*. The university agreed to nominate a member of the governing body and assist with curriculum planning, academic support, guest talks, support for technical projects, undergraduate mentoring programs and access to facilities such as hockey pitches. The DfE gave the go-ahead for UTC Reading to open in September 2013.

Looking back, Rob Wilson stressed the time and effort needed to bring the project to fruition:

> We had to put an enormous amount of effort into the DfE approval process. For example, I personally had to go out and stand on school gates to collect signatures from parents saying they wanted the new school. Overcoming the opposition from local politicians, ideological left wing parents and keeping everyone on track and on mission was a huge task – it was very far from easy.
>
> More broadly, I got the feeling from the Department

[for Education] that they saw UTCs as a sideshow, as something that really wasn't very important as it was only ever going to be a small number of schools and it was only ever going to be an experiment. I never got the feeling that people within the Department believed it could become a widespread movement that was going to work in the long term. They also didn't have the real belief that business could be involved and engaged in schools in the way that UTCs could make them. I just felt that there was a huge dollop of scepticism about the whole idea.

The key to making it work – and the reason why I think it did work – was the support we gained from businesses. Almost every Friday, I was in sessions with local businesses to try and convince them. Not all of them were interested, but eventually we got the right weight of numbers. I opened the door and then the team came in and followed through. It was really intensive, but the fact we had so many strong businesses willing to back the school was a really important part of the process. With them promoting and endorsing it, everybody could see this school wasn't going to fail.
(Wilson R., 2018)

FROM START-UP TO MATURITY

Joanne Harper was appointed principal-designate in September 2012 and took up her post the following January. By then, plans had already been made for renovating the premises, leading to a transformation from an often dark and gloomy building to one filled with light and colour. She also inherited an agreed education plan, but was given a free rein to develop technical projects with local employers. She also led the recruitment of staff and students.

Jo Harper phoned local secondary head teachers individually. While they were not openly obstructive, they did little to support student recruitment and indeed it took another year before she was invited to regular meetings of local head teachers. In the circumstances, other tactics were needed to ensure parents and prospective students heard about the UTC. She said:

We never said no to an opportunity to spread the word. We went to fairs and fetes, primary schools and hosted geek events in pubs. I've never worked such strange hours – and often at short notice! That said, the standout moment was an open event hosted by Microsoft: 300 people turned up, compared with 10 to 12 at other events. This showed the importance of industrial partnerships.

We made sure applicants felt part of the family straightaway, meeting them early and keeping in touch all the way to autumn 2013. We were very pleased with the conversion rate.

We delayed opening the UTC by a week to prepare staff. The first training day was run jointly with industry partners to explain how and why the UTC is different. From day two onwards, we covered essential topics such as safeguarding, fire drills and special educational needs, as well as the technical curriculum.

(Harper, 2018)

The UTC opened with 53 students in year 10 and 82 in year 12. Around 90 per cent were male and 10 per cent female. Year 10 students were asked to take cognitive abilities tests (CAT tests for short), which showed them to be – on average – slightly ahead of expected levels.

An example of an early employer-linked project involved designing a new railway station:

Your brief is to create a new gateway out of Reading Station into the Town Centre, which must:

- Preserve the heritage of the nearby listed buildings
- Be in keeping with the proposed future developments to the southwest and north of the railway
- Provide public amenity (e.g. retail and leisure space) and links to the Town Centre
- Maintain access routes to the Station
- Feel safe and welcoming
- Be sustainable

> All teams will need to showcase their proposed designs. Every team will be interviewed by representatives of the client team, and four teams will be shortlisted to present their designs at Peter Brett Associates' offices. They will do a 15-minute presentation and then have 30 minutes of questions on their design by the Client Team.
> (UTC Reading, 2013)

The UTC was inspected by Ofsted in May 2014. At that time, there were 105 students in years 10 and 11, and 150 in the sixth form. Inspectors graded the UTC outstanding in all categories, praising inspiring leadership, a culture of high expectations, the good progress made by all students including those with a disability or special educational needs, exemplary standards of behaviour and the business-like ethos permeating all aspects of learning.

By 2018, UTC Reading was one of four UTCs in the Activate Learning Education Trust, alongside Oxfordshire, Heathrow and Swindon. In the 2018-19 academic year, there were 453 students on roll.

Not surprisingly, the curriculum evolved over the first five years. In the construction engineering domain, the UTC switched from a key stage 4 qualification in 'design, engineer, construct' to an alternative, 'building information management'. Changes were also made to project-based learning, moving away from year-long projects aimed at the entire school to smaller-scale, shorter projects linked to specific qualification units. Experience showed that running projects over a single term made it easier for employers to be fully involved: they support the launch, visit the UTC twice per half term to look at work in progress and offer advice and visit again at the end of term to review the finished projects.

Michael Halliday, head of employer engagement strategy at UTC Reading, stressed that project-based learning was only one part of the UTC's employer engagement strategy:

> We have probably 200 engagements across the course of the year and most of them are not explicitly curriculum-linked. For example, Thames Water uses a software package called Salesforce. Based on that, the software developers have given us free copies to use

across all the UTCs in the Activate Learning Trust. It's good for their corporate social responsibility and helps young people gain skills in industry-standard software. Similarly, students can learn about cyber security as an extra-curricular activity.

All this means we need better internal communications so everyone knows what everyone else is signed up to in terms of employer engagement. We now produce a spreadsheet identifying every activity on a timeline which is updated weekly. Let's say a teacher invites a professor to come into speak: he or she fills in the spreadsheet which is automatically shared with the team. It's a standing agenda item at all meetings of the senior leadership team to review employer links. And for people outside the SLT, we offer responsibility and autonomy for taking on additional responsibilities involving employer engagement and enrichment activities.

It's actually rare for teachers to know how to link their subject to the real world. We have a checklist for preparing and hosting external links. Has the teacher checked that the employer has accepted the invitation? Will need the employer at reception? Has the employer being briefed before the session starts? It's important for that member of staff to 'own' the link, to know what to do and to buy into it. Then we get it right. Students lap it up because the industry partner is a real person, with real credibility. They answer the question, 'why am I learning this?'

(Halliday, 2018)

Some UTC projects have a competitive element. Being part of the Activate Learning Education Trust has taken this a step further, with teams from member colleges and UTCs competing with each other. A prime example is the Craftsman Cup, organized by the Royal Electrical and Mechanical Engineers (REME). Post-16 students taking part in the challenge start by visiting REME's base at MOD Lyneham, where they explore tanks and operate simulated remote weapons systems. They gain an insight into maintaining and repairing high-tech equipment before engineering solutions to one of 30 real-life scenarios and engineering challenges set for them by REME. The challenge runs from September to May and during that time, REME visits each UTC to provide feedback and encouragement. They can explain why a solution might not work in the real world, how to apply mathematical principles to their work, and offer advice on the best way to present conclusions.

Researchers from the National Foundation for Educational Research (NFER) wrote about the UTC's links with REME:

> The Royal Electrical and Mechanical Engineers (REME), a corps of the British Army that maintains the equipment that the Army uses, works with [Reading] UTC's Engineering Department to deliver a unit of the Level 3 BTEC Diploma in Engineering...
>
> The employer described the project life cycle as having three phases:
>
> 1. The 'understanding' stage where the young people are briefed on the project at a launch day at the REME site where they can gain an understanding of the organisation and the problem that underpins the project. The fact that the project encompasses real-life problems requiring innovative solutions is critical, for example an enhancement to a small tank to enable the secure location of a radiator. The REME interviewee explained: "The project is authentic because it is a real-life project which can make a difference in reality".
> 2. The 'support' stage where the young people return to the UTC, decide on the particular project they want to work on, are allocated to teams and start to work on solutions to the problems. They are supported by REME who supply engineers to mentor and advise young people, four times over the course of the project, as the young people develop their solutions. The young people may also return to the REME site to, for example, take additional measurements or photos.
> 3. The 'recognition' stage where a small panel of military staff will go to the UTC and be

briefed on how the projects have progressed and how the enhancements to equipment work. The young people will present their work and take staff through an exhibition. The panel will grade the projects and the winning two teams will progress to a final stage where their projects will be appraised alongside those from other colleges. UTC staff will consider the projects according to the BTEC assessment criteria.
(McCrone et al, 2019)

In April 2018, a group of year 13 students at UTC Reading spoke about the Craftsman Cup (Harbourne, 2018b). Positive comments included:

- The first task is to set out plans for solving the problem. Then we had to evaluate possible pitfalls and how to overcome them. We had to have fall-back plans, too.
- The biggest value is in preparing skills like teamwork the projects I'll be doing at University.
- It's an exact fit with one of the BTEC engineering units.
- We learnt use a GANTT chart to track progress towards deadlines.
- It improved our communication skills, especially within teams, because you have to discuss how to solve problems.
- A lot of study is individual – working on your own. With this, you're working with others.
- It's good for leadership skills, too. You have to encourage and motivate and keep everyone going, especially if not everyone is equally committed.
- It's useful for time management – to know things have to be done by a deadline.
- We learnt specific skills – for example, we used [software packages] Solidworks and Autocad to design solutions.
- Creating a product that will meet a set of customer requirements – that's a useful employability skill.
- Telling future employers we already have this experience will set us apart from other job applicants.

Negative comments were fewer:

- I did a similar project before, so this feels a bit repetitive.
- It's been very time-consuming: we have to spend extra time outside the timetable to work on this.
- We spent a lot of time designing a solution, but had only limited time to implement it.
- It was spread over a very long period of time, making for breaks in continuity. For example, we didn't work on the project at all in January because we were preparing for an exam in another unit. It might be better delivered in a shorter, more concentrated way. That said, I realize that in real life, it might also be spread over a long period!
- Because this came on top of other coursework, we had a lot of work going on at peak times.

Andrew Thacker, an engineering teacher at UTC Sheffield, said the project brought many benefits:

> For a start, it's real: students are being asked to solve problems actually faced by REME personnel. They have to meet deadlines and achieve goals. They have to work as a team. They gain experience presenting themselves to outsiders, which is very important for their future. They are able to link what they learn in the classroom to the outside world. And of course, they develop practical skills in the process.
> (Thacker, 2018)

Post-16 students at UTC Reading had a variety of destinations in mind. At the end of the 2018-19 academic year, 34 per cent of year 13 leavers went directly into higher education; 18 per cent accepted apprenticeships; 20 per cent went into employment; and 17 per cent went into other forms of education (information supplied by UTC Reading).

Reading University continued to provide support as a partner rather than sponsor. The Pro Vice Chancellor, Professor Gavin Brooks, maintained a close interest:

> When the UTC started, Reading University saw it as part of an outreach programme, our role being to provide opportunities for students to flourish and have

ambitions to progress to higher education. Relatively small numbers have now progressed to this university, mainly onto computer science and biological science courses. Overall, there has been a greater emphasis on apprenticeships than we originally expected, but there is still a good record of progression directly to higher education.

Some of our undergraduates get involved with the UTC either as student mentors or supporting science classes. It's good experience for the undergraduates, some of whom might go on to become teachers, but it is as much to do with volunteering in general and the value that experience adds to their CVs. Our undergraduates also tell UTC students what university life is like and provide advice on UCAS applications. For their part, university staff get involved in curriculum planning, pedagogical support, guest talks, support for practical projects and events for students and parents.

We've been proud to host graduation events for students leaving the UTC. Students who speak at those events are typically highly articulate, highly motivated, engage well with their audience and can explain clearly why they chose the UTC and what they got out of it. A few said they were not doing brilliantly at their previous schools and that the UTC helped develop their potential. There's also a really strong sense of community and identity at the UTC. It's been a very fruitful relationship, and I'm proud to be involved.
(Brooks, 2018)

Rob Wilson believed that attitudes towards the UTC improved over time:

Attitudes towards the UTC changed once people saw it was working. Local councillors who had written blogs opposing the project now came here to cut ribbons – well, that's politics! I'm just grateful that everybody now sees it as an important part of the educational infrastructure in the area.

It's amazing the number of people who watched us open and said, 'I knew it was going to be a good school because it had Cisco, or Network Rail' – brands they knew, that they thought were positive. But it's not just the branding. You also need people within those companies who are dedicated to the success of the school, who are prepared to commit time, energy and thought to make it work.

Whether it's setting up the curriculum or some exchanges into the schools or the business, all that needs time and energy. And when you're doing a day job, that's not always easy. Fortunately most of the people I approached were willing to do that. They came to the monthly meetings and met other businesses to convince them too. They put their name to things and they did everything that was asked of them. If you've got that level of support and energy then it's very hard for other schools and local authorities to hold things back.

Not only that. Once the UTC was open, there was a clear realization from schools in the area that they had to up their game – because they didn't have the bright, shiny, new building for one, but also all these businesses that were backing the UTC: that made their jaws drop! It really did make them think. How do we get those sorts of businesses involved in our school? How do we compete with that sort of offering?

Same goes for Reading College, too. They saw what can be done: they took some of the experience and learning that they gained from the UTC and moved it into the college, and it is a much better college than it was a few years ago.

(Wilson, R., 2018)

12

TENSIONS AND CHALLENGES

Lord Baker was keen to maintain momentum. The main challenge was to secure sufficient funding, particularly the initial capital costs of building new UTCs. Drawing on experience in the health service, he suggested using private sector capital to fund new UTC buildings and equipment. Local Improvement Finance Trusts, launched in 2001, were a means of financing and procuring new primary care facilities. The Department of Health provided start-up funding of £195 million and aimed to raise up to £1 billion of private investment between 2000 and 2010. By 2002, 42 local schemes had been approved, with a collective value of over £700 million.

In the end, HM Treasury and the Department for Education rejected the idea, chiefly because of questions about value for money and the future ownership of school buildings.

This was not the only sign of growing tensions. Michael Gove strongly believed that the key stage 4 curriculum should be based on the English Baccalaureate, or EBacc: English, maths, science, a foreign language and either history or geography. This did not sit well with the UTC curriculum.

BDT's trustees asked Sir Mike Tomlinson to develop an outline proposal for a 'Professional Tec Bac', which would be offered as an alternative to the EBacc in UTCs and other schools specializing in technical and vocational education. His proposal was sent to the Secretary of State, Michael Gove, in February 2012:

A Professional Tec Bac would complement the English Bac by offering a qualification to students who wish to receive a technical education either in a University Technical College or any other school or college which was able to meet its demands.

Such a qualification would address three issues:

1. The need to raise the standard of general education alongside the need to raise the standard of technical education to the level of the UK's international competitors.
2. The shortage of highly skilled technicians up to and including graduate level.
3. The imperative of growing the engineering and technical elements of the UK's GDP.

The Professional Tec Bac would consist of a demanding education with high quality, and rigorous technical qualifications at its heart. We hope that the achievement of this Bac would be shown alongside the English Bac in school and college performance tables.
(Baker Dearing Educational Trust, 2012)

The Secretary of State did not reply in writing, but let it be known that he would not support the Professional Tec Bac.

Instead, he continued to press for a greater focus on rigorous academic learning in key stage 4, while also limiting the recognition of technical and vocational qualifications in published league tables.

Lord Baker commented on the government's latest ideas in his personal diary:

> Woke up to a story in the *Daily Mail* that Gove will recommend the replacement of GCSEs by the old O-levels, with the introduction of a two-tier lower exam similar to the old CSE. This is clearly the work of Gove's eccentric special advisor, Dominic Cummings, as it does not draw upon any educational experience, but a whole pile of prejudice and hostility to the existing educational system.

> *The Independent* asked me to write a piece and I used the opportunity to say that Gove had really shirked the most fundamental questions, that when the [education or training] leaving age rises to 17 next year, and two years later to 18, why should we have an exam at 16 at all? What you need is an exam at 14 for youngsters to then decide which pathway they wish to follow: academic or technical creative or vocational.
> (Baker, K., 2012)

The Secretary of State put it rather differently:

> The introduction of the English baccalaureate measure has resulted in the numbers studying physics, chemistry, biology, history, geography and foreign languages all rising. At the same time, we have already made GCSEs more rigorous by tackling the re-sit culture, ending modules and restoring marks for spelling, punctuation and grammar, but the evidence we have heard from parents, pupils, our best schools and our top universities shows that we need to consider going further...
>
> We would like to see every student in this country able to take world-class qualifications, such as the rigorous and respected exams taken by Singapore's students, for example. We want to tackle the culture of competitive dumbing down by ensuring that exam boards cannot compete with each other on the basis of how easy their exams are. We want a curriculum that prepares all students for success, at 16 and beyond, by broadening what is taught in our schools and then improving how it is assessed.
> (HC Deb 21 June 2012, col. 1025)

Later in the year, the Secretary of State consulted on plans to replace GCSEs with English Baccalaureate Qualifications. There was a great deal of opposition to the idea, and it was withdrawn. The episode nevertheless underlined the importance which Michael Gove attached to academic subjects.

There were also difficult discussions about the future of the

Diploma. Although awarding organizations announced that they would no longer offer the full Diploma after September 2012, they promised to offer the main element – Principal Learning – as long as there was a demand for it. However, DfE said that each stand-alone level 2 Principal Learning Qualification (PLQ) would be treated as equivalent to a single GCSE in the key stage 4 performance tables, even though they took as much curriculum time as three and a half GCSEs.

Professor Matthew Harrison (Royal Academy of Engineering) worked with teachers, engineers and other experts to show how the content of the engineering PLQ could be split into a suite of four discrete qualifications, each equivalent in size to a single GCSE. Two awarding organizations – Pearson and OCR – developed new suites of engineering qualifications, drawing on Harrison's proposals. In addition, WJEC/CBAC (formerly the Welsh Joint Education Committee) developed a suite of qualifications based on the PLQ in construction and the built environment.

Outside the UTC movement, few people mourned the Diploma. It was large, complex and hard to deliver. However, it fitted the UTC model very well; indeed, UTCs had been developed with Diplomas in mind. The government's stance also upset employers who had committed time and energy to developing Diplomas – none more so than business leaders involved in the engineering Diploma, who argued strenuously, but unsuccessfully, for Diplomas to be made a special case.

David Bell, chair of governors of The JCB Academy, had been taken part in preparations for the engineering Diploma:

> As a company [JCB], we were involved in the development of the engineering Diploma, alongside all the big names in engineering manufacturing.
>
> The [coalition] government was wrong to withdraw support for Diplomas – or at least, support for the engineering Diploma. There are always exceptions and they should have had the guts to acknowledge that. I know from long experience in business that can you come up with an idea and it doesn't work everywhere, but you have to make exceptions where it does. The engineering Diploma was really good.
>
> (Bell, 2018)

WEAK RECRUITMENT LEADS TO THE CLOSURE OF THREE UTCS

Two more UTCs opened in 2012, in Hackney and Houghton Regis (Central Bedfordshire UTC). Alarm bells rang even before they opened, with recruitment proving immensely challenging at both UTCs, and also at Black Country UTC.

HACKNEY UTC

Hackney UTC was set up by Hackney Community College with support from the London Development Agency, Homerton University Teaching Hospital and BT. Its specialist curriculum focused on digital media and the health sector. There was reason to hope that students would be excited by links with the so-called 'Silicon Roundabout' – a cluster of digital media businesses in the vicinity of Old Street – while the local area was predicted to create large numbers of roles in para-professional employment in the National Health Service and related sectors.

The original expression of interest for a UTC in Hackney included options for Diplomas, Young Apprenticeships in key stage 4 and Advanced Apprenticeships at 16. By the time the UTC opened in 2012, it was already apparent that none of these opportunities could be offered.

The UTC opened in brand new premises on the Shoreditch Campus of Hackney Community College with 85 students in year 10 and none in year 12. Furthermore, a disproportionate number of the year 10 students had a history of poor behaviour and there were five permanent exclusions in the first year. There were also difficulties with the technical curriculum, as employer engagement got off to a slow start. Relatively high staff turnover did not help, either.

Ofsted inspected Hackney UTC in January 2014, when there were 118 students on roll, all in key stage 4. Two thirds (67 per cent) of students qualified for the government's pupil premium, compared with 28.6 per cent nationally.

Inspectors reported that the UTC required improvement in all areas (Ofsted 2014a). Shortly after the inspection, the founding principal left. When Ofsted carried out a monitoring inspection in May, they found that the acting principal and senior leaders had responded positively and that the UTC's plans were 'ambitious and well-focussed' (Ofsted 2014c, p.2).

However, it was already too late. Only 29 young people applied for places in September 2014, and it was announced that the UTC would close in July 2015. In hindsight, it was clear that there was very little demand for a UTC in Hackney, where education standards were already high; there was no shortage of places in mainstream schools, and little interest in changing school at 14.

CENTRAL BEDFORDSHIRE UTC

Central Bedfordshire UTC occupied two buildings previously used by a school and a college, close to the M1 and A5 in Houghton Regis, north of Luton and Dunstable. Its chosen specialisms were design, engineering and manufacturing. The lead sponsor was Central Bedfordshire College, and the expression of interest was supported by Cranfield University, British Aerospace, the University of Bedfordshire and Central Bedfordshire Council. A firm of consulting engineers and scientists, RWDI Anemos, came on board as a co-sponsor prior to opening. As lead sponsor, Central Bedfordshire College had considerable experience of providing vocational opportunities for 14-16 year olds as well as post-16 students.

It proved impossible to recruit a viable year 10 cohort in 2012 and the UTC was permitted to open with an intake limited to 60 post-16 students. One consequence was that the UTC would be expected to repay some of its first year funding allocation. It was clearly vital to boost recruitment in 2013 and marketing support was secured as early as October 2012. Other options considered (but not pursued) included recruiting at 13 rather than 14, because that was a natural transfer age between middle and high schools in central Bedfordshire (though not Luton), and establishing an apprentice centre adjacent to the UTC. The principal resigned in February 2013.

Ofsted inspected Central Bedfordshire UTC in March 2014, when there were 38 students in year 10 and 58 in the sixth form. While noting recent improvements in several areas, inspectors concluded that overall effectiveness was inadequate and the UTC was placed in special measures (Ofsted 2014b).

A new lead sponsor, Bedford College, took over in September

2014. In the period leading to the arrival of a new, permanent principal, two people shared the role of interim principal. Taking stock of low numbers, it was agreed that the UTC should not recruit to year 10 in 2014.

It is important to note that the UTC's students typically gained distinctions in their engineering qualifications; indeed, post-16 performance tables placed the UTC eleventh in the national table for vocational added value at the end of the 2014-15 academic year. However, student numbers remained stubbornly low: in 2015-16 there were only 101 students on roll. The UTC closed in July 2016 and the site became an outstation of Bedford College, specializing in manufacturing engineering, robotics and automation.

BLACK COUNTRY UTC

Recruitment remained a serious challenge at Black Country UTC. By the start of the 2013-14 academic year, overall student numbers had fallen to 152 across all year groups. The proportion of female students had also gone down; now there were seven in key stage 4 compared with 44 male students. Post-16, there were 17 female and 84 male students. Although a handful of additional students enrolled later in the year, the prospects were not good: the UTC had been funded for more students than were recruited, and the excess would have to be repaid to the ESFA. Staff turnover was high, and when the principal left in February 2014, the vacancy was left unfilled until the autumn.

Ofsted carried out a full inspection in March 2015, a few months after the appointment of a new principal. Inspectors graded the UTC inadequate in every area and placed it in special measures. By now, there were only 158 students on roll against a capacity of 600. Ofsted reported that students were under-achieving, teachers' expectations were too low, poor behaviour had taken hold in classrooms and corridors, attendance was well below average and governors were failing to hold UTC leaders to account (Ofsted 2015a).

The following month, governors announced that the UTC would close on 31 August. Students yet to complete their courses were helped to transfer to other schools, colleges and UTCs. Siemens provided financial assistance to support students' extra transport costs.

BDT conducted an internal review of the reasons for the

failure of the Black Country UTC. These included stubbornly low student numbers, which put the UTC under intense financial pressure; high levels of staff turnover; and a delay in appointing a new principal after Chris Hilton left. There were also changes in membership of the governing body at a time when continuity of leadership would have been desirable. Without criticizing governors who had gone to great lengths to turn Black Country UTC round, BDT came to the view that all UTCs should have one or more governors with direct experience of managing and improving secondary schools.

Then there was the issue of premises. The school which previously occupied the site, Sneyd Community School, was not as highly regarded as some other schools in Walsall, was suffering falling pupil numbers, and in February 2010 was placed in special measures after an Ofsted inspection. Some of Sneyd's history appeared to rub off on the UTC, which found itself coping with an unexpectedly high number of challenging pupils. It was also well away from Walsall town centre, in the middle of a housing estate. And despite having state-of-the-art technical equipment, the UTC simply did not look the part, instead seeming to have taken up temporary residence in one corner of a large and outdated former secondary school.

These drawbacks were identified early and plans were, in fact, prepared to tackle them. One idea would have seen a second UTC – West Midlands Construction UTC – sharing the former Sneyd Community School premises with Black Country UTC. It is fair to say that none of the main parties to this plan were keen, and it later gave way to plan B: brand new premises. This would have enabled Black Country UTC to relocate to the centre of Walsall, where the quality and location of the building would have helped shake off the association with Sneyd Community School. After the 2015 Ofsted inspection, however, it was clear the UTC would not survive long enough for this to happen. Black Country UTC closed, and West Midlands Construction UTC found a site in Wolverhampton.

STUDENT RECRUITMENT PROVED CHALLENGING

By the start of the 2014-15 academic year, 30 UTCs were open, although – as noted above – Black Country and Hackney UTCs were due to close that summer. Student recruitment was a concern for the majority of UTCs, with few meeting their original targets (see Table 1, below).

Table 1: student numbers by UTC, 2014-15 academic year

UTC	Year opened	Full capacity	Students on roll
Aston University Engineering Academy	2012	600	378
Black Country University Technical College	2011	480	158
Bristol Technology and Engineering Academy	2013	484	349
Buckinghamshire University Technical College	2013	600	149
Daventry UTC	2013	600	170
Elutec	2014	600	178
Energy Coast UTC	2014	560	136
Hackney UTC	2012	400	45
Heathrow Aviation Engineering UTC	2014	600	115
Lincoln UTC	2014	640	139
Liverpool Life Sciences UTC	2013	800	445
Royal Greenwich UTC	2013	600	376
Silverstone UTC	2013	576	375
Sir Charles Kao UTC	2014	500	131
The Elstree UTC	2013	600	416
The Greater Manchester Sustainable Engineering UTC	2014	600	105
The JCB Academy	2010	540	524
The Leigh UTC	2014	600	133
The Watford UTC	2014	600	102
Tottenham UTC	2014	924	77
University Technical College Cambridge	2014	670	174
University Technical College Lancashire	2013	800	125
University Technical College Norfolk	2014	600	220

UTC Central Bedfordshire	2012	600	96
UTC Plymouth	2013	650	196
UTC Reading	2013	600	248
UTC Sheffield	2013	600	439
UTC Swindon	2014	600	88
Wigan UTC Academy	2013	500	61
WMG Academy for Young Engineers (Coventry)	2014	640	230

(Baker Dearing Educational Trust, 2014b)

EXAM RESULTS AND STUDENT DESTINATIONS

Black Country UTC received its first full set of results from exams in GCESs, A-levels and Diplomas in 2013. Department for Education school performance tables, published in 2014, revealed that the UTC's results were below average. Half of the students made the expected level of progress in maths, and 61 per cent in English. Thirty-two post-16 students received results. Almost all of them (27) were classed as vocational students; on average, they achieved a grade of Merit- in their qualifications – not a good result (Department for Education, 2014). That said, information about the destinations of year 13 leavers told a rather better story. At the time they left the UTC, all 32 had offers of jobs, apprenticeships or places at a university or college of further education. By the following spring, the DfE reported that 34 per cent were in apprenticeships, 31 per cent were attending a higher education institution and 25 per cent had places at an FE college. A further 9 per cent were in employment with training. The proportion recorded as NEET – not in education, employment or training – was zero (Department for Education, 2016a).

When Ofsted inspected Black Country UTC in March 2015, inspectors concluded that:

> Students' achievement is inadequate because the standards they achieve are well below the national average and should be higher (Ofsted, 2015a, p. 1)

> In 2014, only 21% of students achieved 5 or more GCSEs at A* to C grades. This is well below the

national average and was considerably less than leaders had predicted. No disadvantaged students achieved this measure (Ofsted, 2015a, p. 7)

Achievement in the sixth form is inadequate (Ofsted, 2015a, p. 8)

Unfortunately, Black Country UTC was not the only UTC with weak exam results. In 2015, 14 UTCs sent BDT predicted and actual results for their year 11 students. The percentage predicted to achieve at least five GCEs (or equivalent) at grades A* to C including English and maths ranged from 46 to 78 per cent, with an overall average of 61 per cent across all fourteen UTCs. Actual results were considerably worse than expected. At the UTC for New Technologies, Daventry, only nine per cent achieved at least five grades A* to C. The best result, 71 per cent, was achieved at UTC Reading. Across all UTCs reporting results to BDT, the average was 38 per cent (Baker Dearing Educational Trust, 2015b).

When BDT analysed these results, they found that many more students than expected achieved A* to C in *either* English *or* maths, but not in both. This had a direct impact on the government's headline measure of five or more A* to C grades including both English and maths. Nevertheless, it was clear that something had to be done to improve year 11 exam results, especially in English and maths.

Post-16 results were better. The A-level pass rate across all UTCs reporting to BDT was 90 per cent. Other qualifications taken by UTC students included BTECs, OCR Cambridge Technicals and the Principal Learning Qualification which had previously formed part of the Diploma. Expressed as an average grade, vocational results varied from Distinction* to Merit- across fifteen UTCs reporting BTEC results (Baker Dearing Educational Trust, 2015b).

BDT also collated information about the destinations of students leaving UTCs in 2015. Among 18-year-olds, nearly a quarter (24 per cent) of leavers started apprenticeships, compared with an average of 8.5 per cent across the country as a whole. A further 42 per cent progressed to higher education, compared with 37 per cent of the cohort nationally (Baker Dearing Educational Trust, 2015b).

13

RESPONDING TO THE CHALLENGES

As news spread that some UTCs were facing difficulties, critics began to ask questions.

In November 2013, the editor of the weekly publication *FE Week*, Nick Linford, commented that 'the UTC project appears to be in trouble' (Linford, 2013) and called for the UTC programme to be paused until there was evidence of strong recruitment and good reports from Ofsted. While that was happening, he said, '14 to 16 year-old resources would be better spent on large, well-established and successful FE colleges' (Linford, 2013).

BDT hosted a meeting of UTC principals in December 2013. They were reasonably optimistic about student numbers in the year ahead, but expressed concern about behaviour and attendance. There was a growing sense that some local authorities and schools saw an opportunity to transfer disruptive and challenging students to UTCs. In some cases, UTCs had been given no information by students' previous schools, only finding out about behavioural challenges later. In addition, some students and parents seemed to see UTCs – incorrectly – as an easy option for non-academic children.

On a more positive note, principals present at the meeting said they found their DfE education advisors helpful in providing external validation of progress. Advisors had focused on baseline assessments of newly-recruited students, and had talked extensively to students, staff, parents and in some cases, employers. All principals reported that their advisors considered the quality of teaching and learning to be good or better.

The announcement in 2014 that Hackney UTC was to close brought a fresh wave of criticism. Lynne Sedgmore, Executive Director of a membership organization for large further education colleges, the 157 Group, wrote that while further education colleges were already geared up to provide technical education for young people, 'impatient politicians' had gone ahead and established 'fragile institutions that are not always fit for purpose' (Sedgmore, 2014). She also objected to 'ideas ... born out of ideology, a limited evidence base and, it seems, an almost wilful determination to prove that FE colleges have failed' (Sedgmore, 2014).

Lynne Sedgmore went on to say that 'the Bedfordshire UTC [was] in such a state that an FE college was brought in to help sort it out' (Sedgmore, 2014). In fact, lead sponsorship of Central Bedfordshire UTC passed from one further education college to another, and FE colleges were lead or co-sponsors of a number of UTCs including Black Country, Hackney, Sheffield, Reading, Cambridge, Norfolk, Bristol and Silverstone.

Nevertheless, it was clear that the UTC programme was not going entirely to plan. In December 2013, the Secretary of State for Education, Michael Gove, gave evidence to the House of Commons Education Committee. Graham Stuart, chair of the committee, asked if he was concerned about UTCs' ability to attract pupils. Michael Gove replied:

> The short answer is yes, but more broadly it is important that we analyse UTC by UTC. The principle of university technical colleges is fantastic. The reality has been expensive. The prospect of success in the future is exciting, but there have been one or two cases where recruitment has not been what we would wish it to be ... If you look at the JCB [Academy], it is oversubscribed and hugely successful, so there is nothing either inherent in the model or its delivery which means that the schools cannot be a roaring success, but we need, as ever, to keep things under review.
> (HC 859, 18 December 2013, q. 154)

Behind the scenes, Michael Gove was much more concerned than he appeared when speaking to the Education Committee. He decided to review expenditure on UTCs and in January 2014 he informed BDT that the grant given to new UTCs for specialist equipment – at that time, £1 million per UTC – was likely to be withdrawn unless it was match-funded by employers. Lord Baker recorded in his personal diary that he and Charles Parker met the Secretary of State to discuss the proposal:

> Charles Parker and I were asked to see Michael Gove about the funding of UTCs. John Nash [Lord Nash, Parliamentary Under-Secretary for the School System] also joined us. Gove said that as employers were major beneficiaries of UTCs they should be asked to contribute to the equipment that UTCs need; it was unfair that among free schools only UTCs are given an extra million pounds to buy the specialist equipment that they need. This was a totally absurd request because the curriculum is 40% technical, 60% academic below 16, and 60% technical, 40% academic above 16. For this one needs a considerable amount of specialist equipment.
>
> We pointed out to Michael that in fact companies already give an enormous amount to UTCs in that they are heavily involved in the preparation of the initial bid. We have discovered from experience that the cost of making an application is about £100,000 which is shared between the employers, the university and in some cases the local authority. After a bid has been approved there is a period of between 1.5 and 2 years in the preparation and planning, building and equipping the UTC and agreeing the curriculum. This involves all the companies in considerable extra expense and again we find that the cost comes out at £500,000 of which the government provides £300,000 and the other sponsors provide £200,000. Then after the UTC is open the companies prepare projects to be taught in the UTC and come in and help with the teaching and mentoring. So there is a considerable contribution, not in cash but in kind and effort.
>
> Michael dismissed this and said that the companies are great beneficiaries ... What he proposed was that the DfE would not make it a condition but he wanted

a phrase put into the advice for the next round of UTC applications to say there would be an 'expectation' that the government would put in £600,000 and there would have to be a matching amount from employers, namely £600,000. We pointed out to him that this would virtually kill the UTC movement because small and medium companies could not possibly provide these funds and the larger ones, who are sponsoring two or three UTCs, would not be prepared to do that on such a scale.

Over the following days Charles and I talked to various industrial supporters of UTCs. Uniformly they said they would not put up cash. They are supporting UTCs not for their own company interest alone but for the basic interest of the British economy and to elevate engineering to a higher status. Sponsors of three UTC applications turned down in the last round but invited to resubmit in the next one all said they would not put up the cash. Several made the point that they [as representatives of their companies] don't have authority to do so and would have to go to their main board or charity committee to see if their company wanted to do this.

(Baker, K., 2014a)

Lord Nash invited Lord Baker to a further meeting on 24 February. Lord Baker wrote in his diary:

He said immediately, 'Let's cut to the chase. We have decided to withdraw the matching arrangement but we will restrict the actual amount that can be spent on equipment to £600,000.'

(Baker, K., 2014b)

LOOKING FOR SOLUTIONS

The challenges faced by Hackney, Central Bedfordshire and Black Country UTCs showed BDT that a different approach was needed. Charles Parker said:

> Originally, we thought we needed project managers who would see UTCs through to opening, when they would need no further support from us. We were wrong. UTCs needed a lot of help from us after opening. We needed people with educational expertise and brought on board people like Liz Sidwell, previously a head teacher and then Schools Commissioner at the Department for Education, who could connect with principals because they had been in schools all their lives.
>
> That said, we didn't have many levers. We had influence, not power. In some cases, it took a long time to persuade governors that things needed to change, especially where they lacked experience of running schools.
>
> (Parker, 2018)

BDT commissioned an external review of lessons learned. It was carried out by Tom Revington, previously a consultant with McKinsey & Company. He wrote:

> Attempting to fix things at open UTCs in difficulty has absorbed disproportionate resources.
>
> Some things are far harder to change after opening than pre-opening, including location, building design, choice of Principal and motivation of sponsors. It is far more efficient to get it right first time, even if it takes longer, more resource or deters some weaker projects from proceeding.
>
> (Revington, 2014)

In Revington's view, this meant ramping up the level of guidance and support given to teams preparing bids to open UTCs, and ensuring that only the strongest applications were approved. Between approval and opening, teams should be given more support on issues ranging from curriculum design to staff recruitment; post-opening, there should be an emphasis on identifying and sharing good practice quickly and effectively as well as taking prompt action to tackle weaknesses.

As noted previously, BDT provided DfE with a view on all

applications to open UTCs, though ultimately the decision rested with the Secretary of State. That said, the term 'University Technical College', the abbreviation 'UTC' and the associated logo could be used only under licence from BDT. In two early cases – Nottingham and Newcastle – sponsors were unable to reach agreement with BDT. As a result, Nottingham University Academy of Science and Technology opened in 2014 as a sponsored academy, not a UTC, as did the Discovery School in Newcastle. BDT could also withdraw a UTC's licence if it fell short of expected standards or diverged from the central tenets and principles of the UTC movement. Where this happened – in Greenwich and Tottenham, for example – it was by mutual agreement, where UTCs were converted into mainstream academies.

Revington also identified wider issues that would need to be tackled through research, marketing and in some cases, lobbying. Recruitment was a major challenge that would need concerted action by BDT, UTCs, DfE and local authorities. Funding was another key issue: once fully operational, UTCs were now funded no differently from mainstream schools, despite extra costs associated with the technical curriculum and the longer school day and academic year. He said detailed evidence should be collected and presented to Ministers and officials to support the case for additional funding.

After receiving the Revington report, BDT decided to:

- work with DfE to refine and strengthen the application process for new UTCs
- take a more significant role in the selection of principals-designate
- provide support for principals pre- and post-opening
- produce guides to good practice in employer engagement and student recruitment
- boost marketing activity nationally and advise individual UTCs on marketing and public relations
- commission additional research on UTC running costs
- identify and share good practice in UTC leadership and management.

RECRUITMENT

One reason why numbers were lower than anticipated was the sheer difficulty of providing full, impartial information to prospective students and their parents and carers. Few schools wanted high-attaining students to consider moving at 14; some did not want them to move at 16 either. While they argued that stability and continuity were in the student's best interests, institutional pressures also came to bear. Every student lost to a school represented a reduction in income. And that was not all: the loss of high-achieving students could affect outcomes at 16 and 18, a school's position in performance tables and potentially, even its Ofsted grade.

It was increasingly apparent that some schools encouraged or discouraged transfer to a UTC on the basis of a student's prior record, not his or her aptitude for a technically-orientated curriculum. Students performing well were quietly discouraged from leaving their existing schools, being told for example that UTCs were a form of alternative provision better suited to less able students, or that UTC students were unlikely to be offered places at university. Conversely, schools encouraged students to move to a UTC if they were making less progress than expected and/or had a record of poor behaviour or attendance: losing students expected to achieve low grades would actually help improve a school's position in the league tables.

UTC principals found themselves in a difficult position. They were sure something was going on, but it was hard for them to speak about it openly and publicly without concrete evidence. They were only willing to speak anonymously:

> In our second year, schools got savvy enough to think, 'we've got an opportunity here – we've got this really tough young lad in year nine and there's an opportunity that we've never had before.' They said, 'go and have a look at the UTC. Academia is not for you, go and do something technical instead.' I hate that stereotype. The technical curriculum at a UTC is academically rigorous. But we had this group [of students] who were sent to us for completely the wrong reasons.
> (Principal A, interview with the author)

> Schools are actively discouraging able and well-behaved kids from coming here. I went to see the head at [a local school] and said, 'you can't keep doing

this, you can't keep bringing kids into rooms with their parents and saying "don't go to the UTC, it's inadequate" – because it's not inadequate. So stop doing it.' His reply was 'I don't know anything about that'. They always do that – deny all knowledge.
(Principal B, interview with the author)

The accountability system does this, doesn't it? 'Get these kids off our roll as quickly as possible so we don't have to worry about their results. Our results will look better.' That's definitely what's happened here. We have a significant proportion of kids with poor previous records, who have been very difficult for others. Some have been permanently excluded from their previous schools and others have had fixed-term exclusions.
(Principal C, interview with the author)

Having been on the other side, I can spot it a mile off. The schools use all sorts of different ways of telling us that 'this kid needs a fresh start,' or 'this kid has always been all right for us but he's got a more vocational style of learning,' 'this kid would have been all right with us but there's three or four other kids in his year group that he doesn't get on with and he needs to go somewhere else.'
(Principal D, interview with the author)

We get cases where schools say to a student, 'we could permanently exclude you, but we won't do that if you agree to go to the UTC instead'. That's breaking all the protocols, but it happened as recently as last week.
(Principal E, interview with the author)

Skewed key stage 4 intakes had a knock-on effect on UTCs' post-16 programmes. The original vision was that students would enrol at 14, complete GCSEs and a level 2 Diploma at 16, and then remain at the UTC for a further two years while studying for a level 3 Diploma and/or A-levels. In practice, up to 60 per cent of the cohort left after completing key stage 4. Given that many students had no particular interest in or commitment to the UTC's chosen technical specialism when they enrolled at 14, it was not surprising that some chose to leave at 16 and study other subjects at other schools and colleges. In addition, a significant number of lower-attaining students were not ready for the UTCs' level 3 programmes at the end of key stage 4 and opted for level 2 courses at their local further education colleges instead.

The upshot was that UTCs had to recruit externally at 16+. Again, they found it hard to reach young people and their parents directly, because local schools placed obstacles in their way. That said, students who joined UTCs at 16+ were more likely than their 14-year-old counterparts to be interested in the technical curriculum, enjoy project-based learning and appreciate links with employers.

BDT advised UTCs on ways to boost recruitment at 14, with a particular focus on correcting misconceptions – for example, notions that UTCs were 'the best place for challenging students' and that 'UTC students don't go to university'.

There were particular concerns about year 10 recruitment at Buckinghamshire, Greenwich, Lancashire (Bolton), Plymouth and Wigan UTCs. Financial support from the Gatsby Charitable Foundation enabled BDT to organize targeted support for these UTCs. Marketing specialists reviewed all aspects of their marketing and student recruitment strategies and helped devise tactical marketing plans to address any gaps, funded in part from the Gatsby grant.

In addition, Anna Pedroza (BDT's Marketing Director), Geoff Ashton (Education Advisor) and Alex Hayes (Principal of UTC Norfolk) led a project to collate good practice and helpful resources relating to recruitment. They then produced a guide to student recruitment covering, for example, understanding target audiences (young people, parents, local primary and secondary schools etc), building a recruitment pipeline, and developing a marketing strategy.

BDT also worked with UTCs on a national campaign to promote UTCs to parents. This was based on BDT's survey of parents whose children were aged 14-18 at the time and attending either mainstream schools or UTCs. Two-thirds (68 per cent) of parents of children at mainstream schools felt they were being prepared for the world of work; the figure was much higher – 85 per cent – among parents of children attending UTCs. Writing for the *Daily Telegraph*, Charles Parker said:

Recent research from the Baker Dearing Educational Trust shows that 80 per cent of parents think the current education system needs to change to reflect 21st century Britain, which suggests they have concerns.

The research surveyed 1,000 parents with teenagers at mainstream schools and their responses were compared with 450 parents whose children attend University Technical Colleges (UTCs), technical schools for 14-18 year olds.

The results found that for two thirds (66 per cent) of parents their biggest fear is that their child will not find a job when they leave education and nearly half (48.1 per cent) of parents said they felt stressed about their child's education ...

However, parents with children at UTCs think differently. 70 per cent believe the UTC has made their child more confident in getting a job. Three quarters of UTC parents also believe their child knows what industry they want to work in compared with just half of parents with children in mainstream school.

(Parker, 2015)

RUNNING COSTS

The shortfall in UTC student numbers had a direct impact on funding. Before opening, each UTC's budget was set on the basis of expected student numbers. If those numbers fell short – and most of them did – the result was what the Education Funding Agency regarded as an over-payment, which had to be repaid in subsequent years. The growing burden of debt was a major headache for several UTCs, which felt obliged to curtail spending across the board, for example by reducing staff numbers. With fewer staff it became increasingly difficult to deliver the full UTC vision, including longer hours, a longer academic year and technical subjects taught in expensive specialist workshops.

BDT commissioned an expert in school funding, Michael Stark, to compare data from two UTCs and a mainstream secondary school. He reported that 'actual expenditure per KS4 pupil at the two UTCs was at least 20% higher, but the gains (more intensive curriculum and workshop/lab tuition and experience) were at least 40% higher, than for a KS4 pupil in a typical "ordinary" school' (Stark, 2015).

Michael Stark recommended the introduction of a key stage 4 technical funding premium, similar to premia already paid to post-16 providers and higher education institutions. He argued that that investing in a 20 per cent premium to fund a technical curriculum from age 14 might actually save public money in the long run: UTC pupils who achieved level 2 technical qualifications at the end of key stage 4 were ready to start full level 3 qualifications at 16, whereas many students leaving mainstream schools had to start a level 2 at 16 and were only ready to progress to level 3 at 18+.

Even at the most successful UTC, the JCB Academy, post-16 income was used to subsidise the key stage 4 curriculum. Michael Stark drew on comparisons with Germany and Austria to reinforce the case for a premium for schools delivering a technical curriculum in key stage 4, concluding that 'the [UTC] network will not be able to deliver the planned richer and more focused curriculum at KS4 without a substantial increase in unit funding' (Stark, 2015).

BDT presented these arguments to DfE and Treasury officials, but to no avail. The government was committed to reducing public borrowing. Boosting funding rates for UTCs was simply not on the agenda at that time.

LEADERSHIP AND MANAGEMENT

BDT was increasingly concerned about leadership and management in certain UTCs. Some governing bodies lacked experience of running schools and were not well placed to challenge principals and senior leaders or interrogate performance data. For their part, some senior leaders had little or no experience of running brand new schools or of developing and implementing contingency plans to cope with unexpected challenges. It did not help that a handful of UTCs had to open in temporary accommodation because of delays completing new buildings.

One of BDT's trustees, Sir Kevin Satchwell, agreed to visit six UTCs to identify good practice. Sir Kevin was supported by Vic

Maher, formerly head of Madeley Academy in Telford. In their first report to BDT, they said that in effective UTCs:

1. Senior leadership regularly monitors progress, identifying underperformance affecting individual students and subjects, leading to rapid support and intervention
2. A climate of continual improvement is established by the Principal, with high expectations confirmed by setting stretching targets for individual subjects
3. Subject leadership is provided by a committed subject 'guru' who knows staff well and continually monitors the work of all students
4. Teachers know their learners and their learning needs; marking is diagnostic, teachers provide good feedback, shortcomings are re-visited and learning is consolidated.
(Satchwell and Maher, 2015)

Vic Maher continued to visit UTCs in the following months before preparing material for a new advisory pamphlet published in 2016.

MINISTERIAL INTEREST

In their manifesto for the 2015 general election, the Conservative Party said:

> We will continue to expand academies, free schools, studio schools and University Technical Colleges. Over the next Parliament, we will open at least 500 new free schools, resulting in 270,000 new school places ... We will ensure there is a University Technical College within reach of every city.
> (Conservative Party, 2015, p.34)

After the election – which was won by the Conservative Party – the Minister of State for Skills, Nick Boles, took an increasing interest in UTCs. He visited JCB Academy and Aston University Engineering Academy and hosted round-table discussions with UTC principals and employers. The head of the Prime Minister's Policy Unit, Camilla Cavendish, also met a number of UTC principals.

Ministers and officials explored a number of options to improve the UTC model. They suggested improving leadership, management and teaching by transferring UTCs into Multi-Academy Trusts (MATs). They also suggested that student recruitment – and by extension, funding – could be improved by recruiting pupils at the age of 11.

BDT was initially reluctant to agree to either suggestion. They argued that once a UTC had become part of a MAT, decisions would be taken by the parent trust, not the sub-committee overseeing the UTC. There was no guarantee that MATs would respect the unique characteristics of UTCs, including the central role of employers and universities in their governance, and the extra weight given to the technical curriculum. As for recruiting at the age of 11, BDT initially maintained its view that young people and their parents were in a better position to choose a technical curriculum at age 13 or 14 than at age 11. BDT's preferred solution was to encourage local secondary schools to over-recruit at age 11 in the clear expectation that some would transfer to a UTC (or other specialist institution) at 14. BDT also suggested opening new free schools for the 4-14 age range. Finally, BDT asked DfE to insist that schools provide parents and students with information about UTCs in their area.

Nick Boles stepped back from requiring all UTCs to join MATs. On 15 October, he wrote to all UTC principals and chairs of governors to set out:

> [a] presumption that UTCs should be part of a strong partnership involving successful secondary schools … This could include, for example, joining up with a multi-academy trust or through strong links with a cluster of local good and outstanding schools.
> (Boles, October 2015)

The Minister also accepted BDT's position on the age of transfer to UTCs:

> Thus far UTCs have been for pupils aged 14-19 and we understand most university and employer sponsors think that this is the right age for a young person to

choose to follow a specialised technical route. We will consider proposals for UTCs starting at an earlier age (for example at 13) provided an affordable case can be made that this will improve pupil recruitment and be of benefit to the educational landscape locally.
(Boles, October 2015)

At this point, Nick Boles did not commit to instructing schools to provide their students and parents with information about UTCs, but he told BDT that he would give the matter further thought.

14

SHEFFIELD'S TWO UTCS

Planning for Sheffield's first University Technical College started in 2010, leading to a formal application a year later. The bid was coordinated by Andrew Cropley, then head of Sheffield College's campus at Dyche Lane and later college principal. The future chair of governors, Richard Wright, recalled the early discussions:

> The first people around the table were Sheffield College, the Chamber of Commerce and Sheffield Hallam University. I was chair of the College at the time as well as Chief Executive of the Chamber. My own background was in manufacturing. I'd always been committed to a more vocational route into the world of work and consistently lobbied for a change in emphasis in education. The kind of skills employers need in Sheffield are not always at graduate level – they need technicians as well.
>
> The Chamber is a very interesting place to work because you get involved with businesses of every size, across every sector and it gives you privileged access to people locally, regionally and nationally. So it was perfectly natural for me to lead on recruiting employers to support the UTC. From day one, we involved them in everything the UTC was going to do from equipment specifications and layout to curriculum design.
>
> (Wright, 2018)

Early employer supporters included Boeing, Tata, Kraft, HSBC, Firth Rixson and South Yorkshire Fire and Rescue Service. Sheffield University also supported the bid.

The site chosen for the UTC was the former Sydney Works, close to the railway station, bus interchange and city centre, and owned by the Regional Development Agency, Yorkshire Forward.

The UTC would have two specialisms: advanced engineering and materials, and creative and digital industries. The curriculum would lead to engineering or creative and digital Diplomas, alongside GCSEs and A-levels.

The application was approved in October 2011, the funding agreement with DfE was signed in July 2012 and the UTC opened in September 2013.

Nick Crew was chosen as the UTC's first principal. He said:

> When I first came to this I was a vice principal. I'd got a track record of working in outstanding schools and people said to me, 'why would you take a risk to go to a UTC when you could have quite easily gone to a good or outstanding school as a traditional head teacher and had a very interesting career?' The truth is, there was just a passion burning inside me. I left school at 16 because it didn't quite tick the box for me at the time. Then I picked up an apprenticeship and saw the value in education.
>
> I went to a talk about the JCB academy at Nottingham University when that was just starting and thought, 'wow – this sounds amazing!' I could see myself working in a UTC, and the opportunity came to move to Sheffield. I went to university here. I thought it was very attractive. It was the first head's job I went for and when I got the job I couldn't believe it. Ever since then I've been on a 100 mile an hour journey! Yes, you have sleepless nights, yes, you have your worries and yes, you think life would be easier doing something else, but I wouldn't have it any other way. I just love everything about it: it's such an exciting environment to work in.
>
> (Crew, 2018)

The UTC aimed to recruit 100 students into year 10 in the year it opened, and actually recruited 109. Post-16 recruitment was less successful, with 60 out of 100 places filled. The following year, year 10 numbers were very similar, but 155 students entered year 12. Including students recruited the previous year, the UTC now had 440 students – still some way short of its planned capacity of 600, but better than some other UTCs.

There were also growing signs that schools discouraged high-achieving students from moving to the UTC. At the same time, the UTC was asked to accept students with no particular motivation to study either engineering or creative and media subjects, including students encouraged (or told) to change school mid-year. Richard Wright commented on this:

> Yes, there is a tension, partly driven by the funding system and partly by the fact that schools don't want to lose capable students. We didn't fight them: we tried to appeal to students and parents by calling on employers to explain why they supported the UTC, giving real-life examples of opportunities open to UTC students. We didn't get it completely right and had more 'difficult' students – sometimes at the last minute – then we might have expected. That said, we had a lot of help from the council. They helped neutralize the strongest opposition and recognised that transfer to the UTC was not the best option for every student referred to us by their previous school. And to be fair, it has got better over time because our reputation has grown. We can point to particular successes such as students winning medals at the WorldSkills UK competition or getting places at Oxford and Cambridge.
> (Wright, 2018)

On the other hand, Alex Reynolds, initially Director of Engineering at the UTC and later principal, singled out Sheffield College for praise:

> The support we had from Sheffield College – and in particular Andrew Cropley – was immense. Andrew

led and directed the project to make sure it was on budget and on time, and he made sure we got the right equipment by getting the right advice and guidance from engineers and being sensible about what we needed.

We weren't competing with one another. Students who go to Sheffield College to do engineering are not the students who come here post-16 – and we both supply apprentices to the AMRC [Advanced Manufacturing Research Centre] and students to the two universities. Engineering in the city has gone from strength to strength.
(Reynolds, 2018)

Nick Crew and his colleagues invested a lot time in developing the first set of employer projects:

We decided that we'd go with fully embedded employer engagement in the learning rather than having lessons on the qualification and doing employer projects outside of that. We were worried that if things got tight, teachers would focus on the qualifications at the expense of employer enrichment. What we found by embedding it in and getting the employers to sit down with us and the exam board to contextualize a project to a particular qualification unit … it's really lifted the learning experience for the student, which has raised the attainment. It's also developed their employability skills and resilience. And I think that is the best thing we ever did.
(Crew, 2018)

Ofsted inspected the UTC in February 2016, finding it to be good in every area. Inspectors reported positively on the strong work ethic and behaviour of students. High-quality teaching resources contributed to students' good technical skills and progress in both academic and technical qualifications. There was still room to improve attendance and punctuality at key stage 4 and stretch students in all areas – particularly literacy and extended writing. Commenting on the technical curriculum, the report said:

> The innovative curriculum enables pupils of all abilities to study a specialist course in engineering or creative and digital media, as well as GCSE subjects ...
>
> Technical vocational qualifications within the study programmes are developed in partnership with major employers, with the result that learners' assignments are assessed against the outcomes of real industrial and commercial projects. For example, learners in engineering create lean manufacturing solutions for a manufacturing company, while creative media learners design software for local games companies.
>
> (Ofsted 2016b, p.6)

Ofsted also commented favourably on student destinations. Most post-16 students went on to higher education, but around a quarter started apprenticeships – far higher than the national average. The UTC used student case studies to tell the UTC story:

> UTC Sheffield student Daniel Pickering is celebrating after gaining top grades to study rocket science at a prestigious Russell Group university ...
>
> UTC Sheffield engineering students have won a medal at a prestigious national competition ... Paul Hayter, 18, and Bradley Ellison, 18, won a bronze medal in the [WorldSkills UK] mechatronics engineering category. They were tasked with solving a range of complex automation and mechatronics problems. This involved creating pneumatic and electro pneumatic circuits, PLC programming and machine optimisation.
>
> UTC Sheffield student Madison is celebrating a successful set of academic and technical exam results ... Now she is going on to study a degree in Film and English at the University of Warwick.
>
> UTC Sheffield student Cameron is celebrating after securing an apprenticeship with a top employer. He gained a Cambridge Technical qualification in Engineering, achieving a double distinction, as well as an A-level in Maths ... Now he is going onto a three-year degree apprenticeship with Inspec Solutions based

in Sheffield, [a company] that provides control and safety solutions to leading UK firms.

Studying at UTC Sheffield has put Tom on course for an international electrical engineering career ... The 18-year-old, from Doncaster, has been selected for a sponsored cadet training programme as a junior electro-technical officer with global energy giant BP.
(UTC Sheffield, 2019)

SHEFFIELD'S SECOND UTC

Plans were being formed for a second UTC in Sheffield even before the first one opened in 2013. Part of the impetus came from plans to develop the former Don Valley Stadium – a site which came to be known as the Olympic Legacy Park.

Sheffield Human Science and Digital Technologies UTC, as it was called in the application, would support employment opportunities in the health, sports and digital technologies sector. The bid attracted considerable support not just from the Chamber of Commerce, the city's universities and Sheffield City Council, but from NHS trusts, large and small businesses in the health and digital technology sectors, professional sports clubs and sport and leisure providers.

The decision was taken to form a multi-academy trust to support both Sheffield UTCs. Nick Crew became Executive Principal of UTC Sheffield Academy Trust. The first UTC was now described as UTC Sheffield City Centre and the second as UTC Sheffield Olympic Legacy Park. Alex Reynolds was promoted to the position of principal at Sheffield City Centre and Sarah Clark was recruited to the equivalent position at Olympic Legacy Park, which opened in 2016.

There was a further major development when UTC Sheffield Academy Trust applied for permission to extend the age range at both UTCs to include year 9 (age 13). The Parliamentary Under Secretary of State for Schools, Lord Nash, gave the go-ahead in July 2017 and the first year 9 students joined in September 2018. At that point, Olympic Legacy Park had just over 400 students on roll and Sheffield City Centre, 463.

WHAT STUDENTS THINK ABOUT SHEFFIELD'S UTCS

GROUP 1

The following comments were made during a group discussion at UTC Sheffield City Centre on 19 April 2018. The group consisted of four year 13 students from the two Sheffield UTCs. M = male student; F = female student.

Why did you decide to come to the UTC?

M1: for the opportunity to do engineering. I like building stuff and doing things with my hands.

F1: we moved house and my mum got a job here [Sheffield city centre]. I needed to change schools, so my mum said 'why don't you come here?' At my previous school I was more interested in humanities though I was interested in science too. Doing engineering in key stage 4, I found out it isn't what I want to do: it was very maths-based. That's why I have moved to Olympic Legacy Park to do health and social care and biology. When I leave I want to specialise in biological sciences.

F2: my other school got rid of the electronics course I wanted to do because not many people signed up for it. Then I heard about the UTC and decided to come here.

M2: I was at another UTC in the south of England so when we moved up here I transferred to this UTC. I took GCSEs in engineering and electronics, fell in love with the process of making things and decided to continue with engineering in year 12.

What's it like at the UTC?

F1: the style of teaching is interconnected. There's normal classroom learning and hands-on work, which is really interesting. I hadn't done a lot of practical work before coming here. In a normal school there's a lot of listening and writing down and it was refreshing to do things differently here.

F2: the teachers here are very helpful but at the same time they give you independence. Where I was before, it was, like, generic teaching – just follow the syllabus – and it was quite boring. Here, we've got a lot more support. You can't use the same teaching style for everyone.

M2: the style was the same at my previous UTC as it is here. In both UTCs there's much more practical work than my friends do in other schools. We'll be able to talk about that when we go for interviews for apprenticeships or University. We'll be able to say, "we've done this, we know what that looks like, or how this machine works". By the way, we mostly had the same machines at my old UTC as we have here. The fundamentals are the same.

M1: I really enjoyed the style of learning here [in KS4]. I felt there was much more trust from staff than in other schools, though we also got help from teachers. If we didn't understand at first, they found another way to explain until we do. And they actually know what they are doing because they've worked in industry before – it makes a big difference because they talk from experience, not just a textbook. It makes it feel more genuine. We also had a lot of contact with businesses like Rolls-Royce and Lavender International. That was really enjoyable because they showed how things are done outside the classroom. I'm now at Olympic Legacy Park where we get a lot – and I mean a lot – of employer engagement in IT and we get kit donated to the UTC. Nearly every unit we do has coursework with an employer aspect linked to it.

M2: I'd vouch for that in engineering too. A lot of businesses coming to talk about testing, materials and so on.

F2: I do sport now [at OLP] and I have to say there's hardly any employer engagement. In year 12, we went to a Sharks [Sheffield Sharks, a professional baseball team] training session, but that was about it really. We had to set up and run a sports day, but we did that ourselves – it didn't have anything to do with employers. Oh yes – I remember – Paul Greaves came in, the trampoline coach. We had interviews with him to practise for future job interviews. But when I was here [city centre] there was lots more. In year 10 we went to the Rolls-Royce factory and there were lots of visits by employers to the UTC.

F1: in health and social care – anatomy and physiology – lots of specialists come in from the hospital to talk about anatomy. And when we've done coursework, people come in to look at it – nurses, people from care homes and so on – and listen to us. They check it over, talk to us and give ideas were other things to include. We've been to Northern General Hospital too, and had talks about the types of work available there. A speech therapist told us about how sounds are made, and which sounds become harder to understand with hearing loss.

How could UTCs be improved?

F1: they should share knowledge and ideas more both within and between [Sheffield] UTCs. Some teachers know stuff that others don't, right down to simple things like if assembly has been cancelled! I'd also encourage opportunities for students to go from one UTC to the other to broaden experience, use specialist facilities and meet other students.

M1: I felt I was treated more like an adult at city centre. We were told it was meant to be like a business environment. It motivated students and there was good mutual respect between teachers and students. At OLP we are treated more like children and told what to do. I'm not sure why it's different, but I wish it wasn't. And I would abolish uniforms for sixth formers.

F1: the city centre building is better designed than OLP. Everything over there is open plan, as if for surveillance. It feels like a panopticon – awkward and noisy.

What do you hope to do next?

F1: I've got an unconditional offer from one university to do microbiology but I'm also quite interested in molecular biology so I haven't quite decided yet.

F2: I've had an offer to do sports technology at Sheffield Hallam. My insurance offer is for biomedical engineering.

M1: I'm planning to do animation and visual effects at university. My goal is to run my own business.

M2: when I started sixth form, I was sure I wanted to go straight to university but now I think an apprenticeship would suit me better. I've already had some interviews. I'll probably move back south. I want a higher or degree apprenticeship. In 10 years' time, I'd like to be leading my own projects in an engineering company – say, using SolidWorks [software] to design and make prototypes. I want management responsibilities, too. By the way, before I started at my previous UTC, I was thinking of psychology, English and history A-levels, so this is a total change of direction. I looked at it in terms of career opportunities. There's more need for engineers than psychologists or English teachers. I got really interested in physics, which led to the focus on engineering and then to apprenticeships. I don't want to have big debts or waste time – I can learn while working, with full support from my employer.

GROUP 2

The following comments were made during a group discussion at UTC Sheffield City Centre on 19 April 2018. The group consisted of ten year 12 students from the two Sheffield UTCs. M = male student; F = female student.

Why did you decide to come to the UTC?

F1: from a very young age I have repaired bikes and cars with my grandfather and I've always wanted to do engineering. A friend came here and told me about it.

F2: my mum booked an interview without telling me! We'd spoken about changing school, but I assumed that would be in Rotherham, where I live, not here. I am the only person from my old school to come here. I've always been interested in musical theatre, including the tech side, and that's what I want to do as a career now.

F3: the RAF gave a presentation on science and engineering careers at my previous school. Then one of my dad's friends mentioned the UTC. I did some research and found out about the specialist equipment and links with employers. I came to an open day and then I applied.

What do you think about the UTC's links with employers?

M1: I am interested in an apprenticeship. Teachers can tell me about them and give advice. The UTC has lots of links with businesses, which is better than just looking at careers websites.

F1: I want a career in the media after getting a degree. My class teacher has given me lots of information about all the different aspects of the media and because he knows my strengths, he can give me good advice. Someone from an advertising agency came and listened to my presentation and was very helpful. He's offered advice any time I need it.

M2: I visited a factory, showed someone some of my CAD work and he gave me ideas and advice on getting into industry.

F2: people from Siemens came and gave us advice on preparing for skills competitions.

What do you think about the style of learning?

M3: the UTC definitely focuses on why you're learning things. Where I was before, they didn't really explain what the courses were for or why we were doing things.

F3: in physics and engineering, you can really see the links with maths. Teachers make connections between the subjects here. I did work experience at Mott MacDonald and was able to use the maths I'd learned here, applying equations in real life.

F1: same in English, too. Like letter writing: why we would use a letter, for example to apply for a job and how and why good writing will help us.

M4: we also learn life skills. Whether you work in retail or engineering, you have to learn how to talk to people and how to cope with an environment that is not a school. You're not taught the social skills in most schools – at least, not in the same way. But you're expected to get a job when you leave school or uni. How can you be expected to jump straight in without any preparation or experience?

How could UTCs be improved?

M1: there's not enough computers.

F2: I would prefer fewer, longer lessons. We'd get more work done. As it is, you can get maybe half the task done in one 50 minute lesson. I'm sure people would make more progress in longer lessons.

What's the best thing about UTCs?

F3: students here know how to make and do things.

M2: we have the building blocks for a great career. The experience teachers have from industry means we get industry-standard skills.

M1: when we go for interviews, our portfolios will show what we can do as well as what we know.

F1: we're building the future!

FORMER STUDENTS

The following comments were made during a discussion with two former students – Hannah and Sam – at UTC Sheffield City Centre on 19 April 2018.

Why did you come to the UTC?

Hannah: I went to a standard comprehensive before. I was told I couldn't do technology subjects at A-level so I chose the UTC for the mix of technical and academic subjects. I did consider two alternatives before coming here.

Sam: I started A-levels at my secondary school – French, English, maths and PE – but failed most of my AS exams. The day before the UTC term started, I came to meet the principal, Mr Reynolds, who reviewed my GCSE results and said I'd fit in. I started the next day. I came out with two distinctions in my technical qualifications plus an A-level in photography and an extended project qualification. The course allowed me to be very creative and practical. In my previous school, I was sitting listening and writing notes. I wasn't engaged.

What are you doing now?

Hannah: I didn't know if I wanted to go to university. The apprenticeship route gives me more options. I have started with A-level 3 CAD technician apprenticeship but my employer would support me all the way to a degree and even a master's. I found out about it from a teacher who emailed information about vacancies. I applied the next day. A full assessment day gave me a chance to find out about the company. I did a CAD module at the UTC so I could talk about it at my interview. Engineers I work with take the time to give me advice and a sense of direction. I go to college one day a week.

Sam: I'm at Sheffield Hallam University. I would like a job in video editing. I already do some freelance work – building a portfolio is a very important way of making links and getting on in the media industry. I also have the option for a one-year placement after my second year.

What do you think about the UTC, looking back?

Sam: the teaching here is very engaging and designed to equip us for work. People get jobs in really good businesses. You get lots of honest, constructive, detailed feedback on your work and how you can improve. I liked the emphasis on managing your own learning here, too – it was good preparation for university.

Hannah: the workshops, demonstrations and even basic stuff like health and safety are all very valuable. In academic teaching, too – for example physics, where we did an optional practical endorsement – you get to embed knowledge by seeing things happen for real and as predicted. In my job, I see lots of links with the physics and engineering I learned at the UTC. You need both practical and technical skills and academic knowledge and ability. You get that by combining a technical qualification and A-levels, but it's not offered by most other schools and colleges. Not only that: I did the core maths qualification, which is like real-world maths – finance, mortgages and so on. I am currently buying a house and that knowledge is proving really useful.

By the way, not many people would think of engineering as an office and computer-based career. They see it as just physically demanding workshop stuff. But even in mechanical engineering, you can end up in front of a monitor, designing. The UTC opened my eyes to all this.

(Harbourne, 2018a)

15

JOBS FOR THE BOYS? TACKLING GENDER STEREOTYPES IN UTCS

In 1984 the Engineering Council and the Equal Opportunities Commission launched a year-long campaign called *Women into Science and Engineering* (WISE). The legacy lived on afterwards in the form of *WISE, the campaign for gender balance in science, technology and engineering*.

WISE monitors UK trends in core STEM employment, defined as occupations in science, engineering and information and communications technology. WISE excludes health occupations from the definition of 'core STEM'. Figures published in December 2019 showed that 1,019,400 women were working in core STEM jobs, representing 24 per cent of the total: in other words, men outnumbered women by three to one.

Table 2 (below) shows that the gender gap was relatively narrow in science professional occupations – 45.7 per cent female, 54.3 per cent male – but very wide in engineering professional occupations, where men outnumbered women by almost nine to one.

Under-representation of women in core STEM occupations both influences, and is in part caused by, the attitudes of children and young people. Archer et al surveyed over 9000 10/11 year olds in England. Over 70 per cent reported enjoying science, but under 17 per cent aspired to careers in science; and within the group expressing the most interest in science, boys outnumbered girls by almost two to one (63 per cent to 37 per cent) (Archer et al, 2013, p.176).

Table 2: women in core STEM occupations, 2019

Percentage of women in ...	%
All core STEM occupations	24.0
Science professional occupations	45.7
Engineering professional occupations	10.3
ICT professional occupations	16.4
Science, engineering and technology managers	14.0
Science and engineering technicians	24.5
IT technicians	20.7

(Source: WISE Campaign, 2019)

There are plenty of other examples of gender differences in STEM education. In 2011, for example, an Ofsted report on Design and Technology (D&T) said that:

> Despite all boys and girls studying D&T from ages 5–14 years, when they have an opportunity to select an area of D&T to continue to study at GCSE level, the choices boys and girls make have been distinct. In 2007, national data shows that girls chose mainly to study food and textiles and were well represented on graphic products courses. Boys chose mainly to study resistant materials, electronics, systems and control, and engineering.
> (Ofsted 2011a, p.53)

In another 2011 report, Ofsted identified differences in boys' and girls' attitudes towards careers, even at primary school:

> Almost to a pupil, the girls spoken to would not countenance actually taking up a role they associated with males, for example plumbing, 'because it is so messy and boys like mess'. Both boys and girls had considerable difficulty coping with the concept of a boy being a nursery nurse, however theoretically possible they admitted that it could be. (Ofsted 2011b, p. 7)

Ofsted also visited secondary schools. They found that girls in single-sex schools were more likely than most to consider jobs stereotypically done by men:

> The girls in the selective secondary schools did not view any career as being unavailable to them, as long as they worked hard and achieved the relevant qualifications. They understood that this was integral to the changing roles of women and they felt that more women should be encouraged into roles traditionally done by men. However, this confident thinking, strongly championed by teachers and school leaders, was not matched by any noticeable shift away from gender-typical course or career choices. Almost all of these girls told inspectors that they were not planning to pursue such a route for themselves.
> (Ofsted 2011b, p. 9)

This is confirmed in other research. Mujtaba and Reiss (2013), for example, examined gender differences in the take-up of physics A-levels in England. They noted that girls were at least as likely as boys to achieve high grades in GCSE physics exams (that is, at age 16). They went on to research the attitudes and perceptions of 15-year-olds who intended to continue to study physics post-16. They reported that 'Such girls had lower confidence in their conceptual ability and lower physics self-concept than such boys even though there was no difference in their conceptual ability' (Mujtaba and Reiss, 2013, p. 2989).

For this and other reasons, boys typically outnumber girls in A-level physics classes. In 2019, there were 34831 A-level physics entries in England, of which 22.4 per cent (7701) were female and 77.6 per cent (27961) male (Joint Council for Qualifications, 2019).

Suffice to say, UTCs were aware from the outset that pre-existing attitudes towards science, technology and engineering would affect the recruitment of female students. JCB Academy decided that in principle, the year 10 intake should comprise equal numbers of boys and girls. In practice, few girls applied; those who did were very likely to be offered places. Conversely, however, there were not enough places available for all the boys who applied. A parent of a boy who was not offered a place argued that the

Academy's admissions policy placed boys at a disadvantage. The Schools Adjudicator (2013) upheld the complaint, concluding that the policy was indirectly discriminatory. This ruled out any form of positive discrimination: instead, UTCs had to appeal to girls and their parents through strong messages and the power of persuasion.

This has only ever been partially successful. In 2014, *Schools Week* extracted figures from the Department for Education's schools census which revealed that only 21 per cent of UTC students were female (Nye and Camden 2014). Figures used in their report are set out in table 3.

Table 3: UTC student numbers, 2014, and proportion (per cent) female

UTC	Female	Male	Total	% Female
Black Country	25	120	145	17%
Central Bedfordshire	5	100	105	5%
Aston University Engineering Academy	35	260	295	12%
Wigan	10	50	60	17%
Hackney	65	50	115	57%
JCB Academy	50	380	430	12%
Reading	15	125	140	11%
Daventry	15	80	95	16%
Elstree	115	115	230	50%
Plymouth	30	110	140	21%
Buckinghamshire	5	85	90	6%
Liverpool	120	60	180	67%
Bristol	15	175	190	8%
Silverstone	20	140	160	13%
Sheffield	25	185	210	12%
Greenwich	55	225	280	20%
Visions Learning Trust	*	70	70	<5%
Total	605	2330	2935	21%

Notes: figures rounded to the nearest 5. *Fewer than 5. (Nye and Camden, 2014)

It is notable that the three UTCs where female students accounted for at least 50 per cent of all students did not specialize

Aston University Engineering Academy

The inaugural Duke of York Awards ceremony
at Buckingham Palace, 2013

The JCB Academy – hydraulics

Black Country UTC – engineering workshop

Lord Dearing

Michael Gove and Lord Baker at Elstree UTC, July 2014

Computing students at UTC Reading

UTC Sheffield City Centre – CNC milling

UTC Olympic Legacy Park – robotics

UTC Northern Lincolnshire – engineering

Justine Greening and Lord Baker at Scarborough UTC, 2017

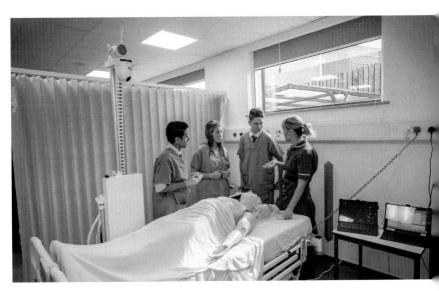

Health Futures UTC

in engineering: instead, they specialized in life sciences, healthcare, digital media production and technical skills for the film, theatre and visual arts industries.

Schools Week spoke to Professor Alison Wolf about the figures. She told their reporter that 'it is going to be difficult to persuade many 14-year-old girls to opt for a boy-dominated specialty and boy-dominated classrooms for the rest of their school lives ... Keeping girls' options open, so they can make choices when they are more mature, strikes me as a better bet' (Nye and Camden 2014).

As it happens, and as the *Schools Week* article acknowledged, BDT was already working with WISE and others to share ideas for improving the recruitment of girls to UTCs. The Royal Academy of Engineering commissioned WISE to write a guide for UTCs, with additional support from the Women's Engineering Society and WiSET at Sheffield Hallam University. *University Technical Colleges: opening up new opportunities for girls* (Best, 2014) was published in June 2014 and distributed to all UTCs that were open or in development. The lead author was Fay Best, who said:

> Our thinking was that if girls didn't apply to UTCs and only boys applied, then rather than improving the situation in terms of gender they might actually make it worse. So it was critical that somebody worked with them ... The Royal Academy of Engineering came on board with us. They agreed that there was an opportunity to make changes, to make a difference, and they sponsored us to write the brochure.
> (Best, 2018)

This was followed by a series of half-day workshops for principals, governors, lead staff and sponsors of UTCs. The guide and workshops explored the barriers to girls and women choosing to pursue science, technology and engineering education and careers and set out strategies for engaging girls and young women in STEM education and careers, including:

- Preparing the ground – for example, naming champions to drive forward diversity objectives
- Understanding the data on female participation in the UTC and the workforce, and setting targets

- Providing unbiased information that appeals to girls
- Engaging with parents, families and careers advisors
- Providing access to female role models
- Developing an inclusive learning environment
- Supporting girls to make a career in STEM
- Rewarding and celebrating success

To back up the WISE workshops, BDT surveyed female students in 12 UTCs (Baker Dearing Educational Trust, 2014a). A total of 159 students chose to take part in the survey, of whom 111 (70 per cent) attended one of three UTCs – Elstree (multimedia arts, performance and production technology), Greenwich (engineering and construction) and Liverpool (life sciences). The results were therefore indicative rather than fully representative of female students' thoughts and attitudes.

The reasons most likely to be given for choosing to attend a UTC included 'facilities and equipment', 'opportunities for work experience' and 'the technical specialism(s)'. Reasons least likely to be given were 'my previous school encouraged me' and 'my friends were attending'. Girls rated their UTCs as better than their previous schools, especially in relation to specialist equipment, links with business and employer projects. A large majority – 87 per cent – enjoyed attending a UTC, while seven per cent did not and six per cent said they were indifferent. Asked about the area they would like to work in as adults, respondents' top choices were arts and culture (21 per cent), engineering (16 per cent), healthcare (16 per cent) and science and research (13 per cent). A majority felt that women could access the same job opportunities as men in key STEM industries, but the level of agreement varied: 83 per cent thought women had the same opportunities as men in science, 79 per cent in product design, 75 per cent in technology, 65 per cent in engineering and 54 per cent in manufacturing (Baker Dearing Educational Trust 2014a).

The WISE guide, workshops and student survey fed into an overall guide to student recruitment published by BDT (Baker Dearing Educational Trust, 2015a). UTCs were strongly advised to have explicit plans for recruiting girls, covering (for example) –

- Activities with primary schools to encourage female students to think about STEM related subjects and careers early on

- Case studies of girls and young women learning at the UTC or involved in STEM-related careers
- Girls and young women acting as spokespeople for the UTC at open days, in the media, and at events
- Imagery that portrayed girls and young women working alongside boys or in small groups together
- Special events aimed at girls: for example, UTC Reading invited girls to open days two hours earlier than boys
- Information about careers that communicated how a role could make a difference – an important consideration for girls
- Offering subjects related to their specialism that might be more likely to interest some girls: for example Greenwich UTC offered architecture as part of the construction specialism.

The guide also included a case study, describing work done by UTC Reading:

> UTC Reading has worked with WISE to look at how they can recruit more girls. They set up a STEM Hub Group and invited the local primary and secondary schools, business STEM ambassadors, professional organizations and some of their female UTC students to join. The group meets once a quarter and has committed to a number of events to recruit girls into IT and Engineering.
>
> Through the STEM hub, they have brought together Fujitsu and CGI to create a Girls IT Challenge that will be run at the UTC and have also worked with Thames Water and Kier to create a Girls Engineering Challenge.
>
> They have invested in the F1 in Primary Schools kits and invited two local primaries to take part in the project. This was a huge success and next year five primaries have been given the opportunity to take part. To qualify each team must include a least one girl. They are also a regional hub for F1 in Secondary Schools and have targeted secondary teams from girls' schools in the area.
>
> Last year they hosted a Lego Mindstorm IT challenge to celebrate International Girls in ICT Day. Six primary schools participated in the all-female

challenge. This engaged students and their teachers and also generated valuable media coverage portraying girls taking a leading role in IT.

Throughout the UTC there are images of young women engaged in STEM related activities and the website includes case studies from female students about why they selected the UTC and what they hope to do in the future.

(Baker Dearing Educational Trust, 2015a)

Some progress was made after 2014. Based on figures given to BDT by 47 UTCs in autumn 2019, the proportion of female students rose by eight percentage points over those five years, from 21 per cent to 29 per cent. Overall, 13,542 students were enrolled in UTCs in 2019, of whom 9572 (71 per cent) were male and 3970 (29 per cent) female.

That said, the percentage of female students varied widely across the UTC network. At Mulberry UTC, specialising in health and creative industries, 80 per cent of students enrolled in 2019 were female. Girls were also in a majority at UTC Bolton (specialist subjects: engineering and health sciences), Elstree UTC (multimedia arts, performance and production technology), Global Academy UTC (broadcast and digital media), Health Futures UTC (healthcare and health science) and Liverpool Life Sciences UTC. At the other end of the spectrum, there were conspicuously low percentages of female students at Heathrow (aviation and engineering – 10 per cent female) and Buckinghamshire UTC (IT and construction – 11 per cent female).

Summing up, Fay Best said:

> Gender needs to be a focus of every principal and their senior team ... but at the moment I don't know if it is.
> There's been a focus on getting UTCs up and running and getting enough pupils in [regardless of gender]. I would like to see that change sooner rather than later because once a UTC becomes a boys' school and is *seen* as a boys' school, it's much harder to change.
> (Best, 2018)

CASE STUDY: LEEDS UTC

Leeds UTC specializes in advanced manufacturing and engineering. When it opened in 2016, girls represented 13 per cent of year 10 students; two years later, the percentage had risen to 23 per cent.

It is not entirely coincidental that Fay Best is a member of Leeds UTC's governing body. She helped colleagues adopt good practice, as identified not only in the original brochure for UTCs, but in later WISE campaigns such as *My Skills My Life* and before that, *People Like Me* – wider programmes to encourage girls throughout the country to consider STEM studies and careers.

In an extended interview, Best identified a number of significant factors. For example, Leeds UTC encourages female students to act as ambassadors, showing visitors round and answering their questions. Best believes they have an important part to play in reassuring girls who are considering transferring to the UTC. Their mothers, too: 'Mothers are key influencers. They meet the ambassadors, who are quite confident young women. They often comment, "gosh, I would like my daughter to be just like her. She looks like she's enjoying herself"' (Best, 2018).

However, success can only be achieved if good practice is embedded in all aspects of the UTC's work, starting with the way engineering, science and technology are depicted and described. Best says that engineering careers are too often illustrated with pictures of men in hi-vis jackets and hard hats, with women very much in a minority. In addition, engineering tends to be described in terms of processes and activities, and much less as a means of solving the world's problems.

Fay Best said:

> We looked at the images used in the UTC prospectus. If a photographer came and took photographs of a classroom they would generally take a photograph from the back of the room, showing the whole class. What you get then is maybe 12 boys and two girls. Girls looking at that image would see a class of boys. All you've got to do is zoom in on that image so that you've got one girl and two boys and a teacher – and maybe the girl in a positive stance, as opposed

to passive – and you've got a much more inclusive picture. Girls will see themselves in a powerful situation and they won't feel isolated.

We also looked at the words we use. Boys generally want to know *what* they're going to be doing. Girls want to know *why*: what the outcomes are. The bigger picture.

For instance, [Leeds UTC] did a project about diluting acid in a local company. Acid is brought in neat, because obviously there's less of it and it's less expensive to transport that way. Then the company adds water to it and it gives off heat. The effect can be calculated and predicted by applying maths and physics principles. But in order to get the girls interested in it I recommended re-writing it so it had a story, linking it to the issues of global warming, the fact that we're using fossil fuels for public transport, and so on. All of a sudden the girls could see *why* they were doing this task and these calculations – how it all links to wider issues. And then, what about capturing and using the heat that was given off? There would be benefits from that, too. It was a case of taking a very practical task, previously described in terms of 'this is what we're going to do', and placing it in the wider context of the environment and benefits to society.

Or benefits to individuals, for that matter. In another project, students had an interview with a young lad that used a prosthetic limb. They then had the opportunity to design and make some prosthetic limbs themselves, looking at the design, the manufacturing process, the practicalities and how they would need to be adapted for individual people. Actually seeing the benefits at the end of the project made the students feel empowered and they went away with a feel-good factor.

But we have to think about language the whole time. At a past visit to a UTC, I asked the principal, why have you put you put 'men' and 'girls' on the toilet doors? That's an obvious mistake, but language is often almost subliminal. It's easy to make girls feel lessened by using just one wrong word.

Unconscious bias plays a huge role. When we [the

WISE campaign] train ambassadors to deliver 'My Skills My Life' outreach events, the first thing we do is an insight into unconscious bias. But it's not just about the ambassadors, or the teachers in a UTC. They might not have a gender bias themselves, but the girl sitting in front of them might have a bias, conscious or unconscious, that engineering's for boys, not girls – and it may well be that her parents feel the same way.

There are a lot of people in UTCs who are very passionate about getting girls into STEM but they do need to take advice and ideas from people who have specific expertise in this area. And none of it costs money really: all the things they need are readily available on the WISE website. If you publish anything, use the right language and the right images. It doesn't cost any more to do so, but it makes it more inclusive.

One rule of thumb is that if something is designed to appeal to young women it will still appeal to young men, but not vice versa. Boys are coming anyway. They know it's for them. They already have an interest in STEM subjects and careers and they can see it's for them. It's the girls that we need to market to. Once you've got both boys and girls in a UTC you will get to a tipping point. The girls will speak for themselves. The results will speak for themselves. But we've got to get it right from the beginning. It's much easier to do it right in the first place than it is to try and change it later.

(Best, 2018)

16

THE MOVEMENT GROWS

Eleven UTCs opened in the 2015-16 academic year: West Midlands Construction UTC (Wolverhampton), UTC Harbourside (Newhaven), UTC Bolton, Derby Manufacturing UTC, Health Futures UTC (West Bromwich), Humber UTC (Scunthorpe), UTCMediaCityUK (Salford), Medway UTC (Chatham), South Devon UTC (Newton Abbott), South Oxfordshire UTC (Didcot) and South Wiltshire UTC (Salisbury).

Richard Garner, education editor of *The Independent*, attended the official opening of Health Futures UTC on 4 February 2016. He was particularly interested in the UTC's technical challenge projects, which included working with 'robot patients' which could exhibit life-like symptoms (Garner, 2016). The UTC aimed to provide a broad introduction to health and social care, leading to careers as doctors, paramedics, nurses and other healthcare professionals. The UTC's lead partners were West Midlands Ambulance Service NHS Foundation Trust, Midcounties Co-operative Pharmacy and the University of Wolverhampton, but there were links with many other NHS trusts across the West Midlands as well.

Three days after Garner's article appeared in *The Independent*, *FE Week* published its latest review of UTC numbers under the headline 'Numbers falling, closing down – University Technical College revolution fails to deliver' (Robertson, 2016), reporting that six of the 15 UTCs opened between 2010 and 2013 had experienced a fall in student numbers in the 2015-16 academic year. Wigan UTC was operating at 14 per cent of its potential capacity, whereas

Sheffield UTC was nearly full and JCB Academy had actually exceeded its planned capacity (Robertson, 2016).

By this time, a decision had been taken to close Central Bedfordshire UTC due to low pupil numbers, and Royal Greenwich UTC was preparing to become a mainstream 11-19 school. Other UTCs were finding it hard to recruit students, which led to further financial pressures and in some cases, increased staff turnover.

The Minister of State for Skills, Nick Boles, wrote to all UTCs again on 3 March. He said he was considering whether local authorities should be required to write to parents of year 9 children about options for study at 14. He wrote, 'I have seen first-hand the difference to pupil recruitment that can be made when local authorities have written to parents and potential pupils directly about their local UTC's offer' (Boles, 2016b).

Accepting the challenges faced by many UTCs, Boles also announced a package of support for UTCs yet to be judged good or outstanding by Ofsted. The package included support from a recognised 'Teaching School' (a high-performing school approved to support school improvement in their local area), an experienced mentor for the UTC principal, support for UTC boards of governors and enhanced education advisor visits in the second term of a new UTC's operations. The Department's education advisors would work with BDT's own education team to ensure UTCs were offered consistent and well-coordinated support.

Finally, Boles took a firmer stance on multi-academy trusts. There would in future be a presumption that new UTCs would be part of either a MAT or a partnership of 'similar depth, strength and permanence', with a clear preference for the former (Boles, 2016b).

In a separate letter to Lord Baker, the Minister re-stated his interest in expanding the age range of UTCs, 'whether through a feeder school or otherwise'. He also wrote that 'the Chancellor, Secretary of State and I are strongly supportive of the unique features of UTC[s] – schools that deliver much needed technical education with a strong role for employers and universities' (Boles, 2016a).

BDT welcomed the Minister's plans. Reservations had not gone away, but were tempered by the fact that some UTCs were already part of MATs. Wigan UTC, for example, had joined the Bright Futures Educational Trust in 2014. Furthermore, a MAT with a UTC in its portfolio, Leigh Academies Trust, had already announced plans to open an 11-14 school alongside its UTC. BDT appreciated that this could become a test-bed for extending the UTC age range.

BDT explored the idea of establishing a new MAT specifically for UTCs, the UTC Network Trust. In July 2016, BDT's board heard that the proposed trust would primarily benefit groups applying to open UTCs in the future, but it was also possible that existing stand-alone UTCs might choose to join. The UTC Network Trust would:

- Rapidly improve educational standards in UTCs
- Drive the agenda for the best possible 14-19 technical education in the system
- [Be] a strong protector of the UTC brand
- Promote the strategic development of the UTC programme
- Drive scale economies within its branches through central financial management
- Share best practice with its own UTCs and also with UTCs outside of UTC Network Trust
- Build a UTC learning community, focussed on teaching in and operating a UTC
(Baker Dearing Educational Trust, 2016)

It was anticipated that the new trust would be a separate charitable company, but with strong links to BDT. The plans were presented to the Department for Education, but a response was not immediately forthcoming.

Meanwhile, the government had not reached firm decisions on the future of the EBacc, even though the consultation period ended in January. Lord Baker was concerned about the proposed target for 90 per cent of young people to take the full set of EBacc GCSEs, not just from the perspective of UTCs, but because of the impact it would have on mainstream schools and their students. Writing in the *Daily Telegraph* on 13 April, he described the EBacc as 'almost word-for-word the curriculum laid down by the Secondary Regulations of 1904' (Baker, K., 2016a). He said:

> All young people should make and do things as part of a broad and balanced curriculum. The digital revolution needs students to be work-ready; by the age of 16 they should have had experience of working in teams, solving problems and making data-based decisions. They should

> acquire basic business knowledge, the skills of critical thinking, active listening, presentation and persuasion. University Technical Colleges aim to do that, but there are just 38 of them open in the country. Though 20 more are being readied, we need to add to that.
>
> (Baker, K., 2016a)

Lord Baker took the journalist, Peter Wilby, to see Lincoln UTC. Wilby said:

> Each college has a university sponsor and several industrial sponsors that provide work experience. The involvement of universities is a stroke of genius. Technical education – engineering, computer science, and so on – has a lamentably low status in England, and although the word "technical" is a turn-off, "university" gets parents' juices flowing …I hold no brief for him, but in these strange times I am tempted to address him as Comrade Baker.
>
> (Wilby, 2016)

A few days later, it was the turn of the *Financial Times* correspondent, Miranda Green. Lord Baker took her to see the UTC in Sheffield city centre. She wrote:

> For Lord Baker, the squeezing out of technical subjects is 'completely crazy' and he laments the lack of focus on skills shortages. At Sheffield's UTC, boys and girls of startling maturity are learning robotics from a technician seconded here by Siemens, in a mini-factory sponsored by Festo. 'I just like making things,' as one student put it.
>
> The subjects may be hard – maths, physics and engineering – but in the lab they have a clear, practical use.
>
> (Green, 2016)

Lord Baker had been talking about the digital revolution for some time. In May, he set out his ideas in an Edge Foundation report:

> I have believed up to now that technical revolutions create more jobs than they destroy ... However, I have now come to a different conclusion and I now think that the Digital Revolution – the Fourth Industrial Revolution – will not follow this pattern.
> (Baker, K., 2016b, p.6)

Lord Baker wrote that the pace of change was faster than ever before; agents of change were proliferating; change could be brought about not just in laboratories, but by individuals in their own homes; and huge amounts of capital were available to develop, implement and market these agents of change. He argued that these trends would inevitably have a major impact on the labour market. Computers could take on tasks requiring high-level reasoning, potentially leading to huge job losses. On the other hand, computers and robots could not easily take on tasks which involved non-routine physical tasks such as social care, or the attributes of empathy and creativity. Summing up, Lord Baker said:

> In other words, knowledge is as necessary as ever, but it is not enough. Abstract knowledge and reasoning need to be connected with the real world through practical applications. A play assumes a certain meaning when read out silently; more when it is read aloud; and more again when it is performed. The same is true of maths, physics, art, music, design, technology ... all come alive when used in a meaningful way.
>
> This is the philosophy behind University Technical Colleges. Employers and teachers collaborate to design real-world projects undertaken by groups of students over a period of weeks. Each project has a direct connection with the world of work and leads to a tangible outcome such as a design, product or presentation – and often all three.
> (Baker, 2016b, p.17)

THE SAINSBURY REPORT

In July 2016, an independent panel on technical education, chaired by Lord Sainsbury, published its report (Department for Education,

2016b). The government responded the same day, accepting the panel's recommendations.

SUMMARY OF MAIN RECOMMENDATIONS

The panel proposed that the vast majority of 16 year olds would choose between two main options:

- an academic option, for those who are aiming to progress to a full-time undergraduate course at university at 18. This option would include A-levels. Reform of this option fell outside the panel's remit
- a technical option, for those wishing to gain the technical knowledge and skills required to progress to skilled employment at 18. There would be two modes of delivery:
    - apprenticeship
    - a two-year, school- or college-based technical education route

Short bridging courses would enable individuals to move between the academic and technical education options in either direction.

The panel proposed 15 technical education routes including construction, creative and design, digital, engineering and manufacturing, and health and science.

Each route would –

- be aligned to apprenticeships
- provide an initial 'common core' for everyone on the chosen route, followed by specialised options to prepare students for entry into a specific occupation or set of occupations
- include –
    - a single set of maths and English requirements
    - short duration work experience placements and/or work tasters in year 1
    - a longer work placement in year 2

Individuals not ready to access a technical education route at age 16 would be offered a transition year to help them to prepare for further study or employment.

Technical education qualifications – later named 'T-levels' – would be based on requirements defined by panels of industry professionals convened by the Institute for Apprenticeships (which the government decided to rename the Institute for Apprenticeships and Technical Education) and aligned with approved apprenticeship standards. T-levels in each technical education route would be offered and awarded by a single body or consortium under a licence awarded after open competition.

T-levels would be based on achievements in the common core for the route and relevant specialist/occupation-specific knowledge and skills. Assessments would include realistic tasks as well as synoptic assessment which, together, would test a student's ability to integrate and apply their knowledge and skills.

The government published an action plan in October 2017 which spelt out T-level plans in further detail (Department for Education, 2017). T-levels would combine:

- technical knowledge and practical skills specific to their chosen industry or occupation
- an industry placement of at least 45 days in their chosen industry or occupation to aid the development of common workplace skills
- relevant maths, English and digital skills

Following the Wolf Report on vocational education (see chapter 6, above), level 3 technical and vocationally-related qualifications approved for use in schools and colleges had been categorised as either 'technical qualifications' or 'applied general qualifications' (AGQs). Technical qualifications and AGQs were offered by a range of competing awarding organizations, including Pearson (BTECs), City and Guilds and OCR. The Sainsbury report and subsequent government consultation papers strongly suggested that T-levels would replace all of them.

However, many schools and colleges opposed the idea of phasing out AGQs, some of which were offered in combination with other qualifications including A-levels. Many UTCs, for example, offered a combination of a technical qualification or AGQ alongside one or more A-levels. BDT was very concerned that the large size of the T-level – 900 guided learning hours – would make it impossible for UTC students to take an A-level – for example, in maths – as well as a T-level. This might limit their chances of studying engineering at university or gaining prestigious higher and degree apprenticeships with employers.

The issue rumbled on for years. In summer 2020, it was still unclear whether AGQs would be phased out or reprieved.

MAKING ENDS MEET

The financial challenge facing UTCs was increasingly obvious. Accumulated deficits ranged from a few hundred thousand pounds to over £1 million.

BDT continued to lobby Department for Education and Treasury officials for a key stage 4 technical premium, building on Michael Stark's evidence that the cost of delivering the full UTC curriculum was 20-30 per cent higher than the cost of the typical curriculum in mainstream secondary schools. BDT argued that a technical premium should be available to any school with a strong focus on technical subjects in key stage 4, echoing long-standing uplifts paid to providers of post-16 technical education. The Treasury took a neutral stance, suggesting that any premium would have to be funded from the DfE's global budget.

Simon Connell, then BDT's Senior Business Advisor, analysed detailed data from a cross-section of UTCs. He found that costs varied significantly, sometimes unexpectedly so: it was, for example, hard to understand significant variations in cleaning and energy costs.

Connell drew on his analysis in a new guide to financial planning in UTCs (Connell, 2016). For start-ups, Connell recommended that new principals should be extremely cautious about building high fixed costs into their projections, because it would be difficult to change direction quickly or cheaply at a later date. As a rule of thumb, the costs of employing teaching staff (salaries, National Insurance, pension contributions etc) should not exceed 60 per cent of projected income after a UTC's first year. This could only be achieved by controlling pupil-teacher ratios, with an ideal of one teacher to every 15 pupils: a ratio worse than around one to 12 would result in financial difficulties.

Connell also noted that some UTCs tried to offer a wide choice of optional GCSE subjects, which of course meant employing additional subject specialists: even if employed part-time, the additional cost would be hard to justify when student numbers were low. Finally, in their early years, Connell believed UTCs should think creatively about other staff costs, for example by sharing costs with each other or with other schools and colleges. He recommended spending not more than 12-14 per cent of UTC budgets on non-teaching staff at first, with the aim of reducing this to 8-9 per cent over time.

There were increasing demands on BDT's time. Originally, it had been expected that BDT's role would focus mainly on advising groups preparing bids to open UTCs. In practice, the work did not stop once a UTC opened its doors: in many cases, it intensified. Among other things, BDT's team provided advice on student recruitment, qualifications and curriculum, finance, and improving the quality of leadership and management.

Where UTCs were performing well, BDT's main aim was to help principals and chairs of governors to share ideas, experience and good practice through regional meetings and an annual national conference. Each new UTC could expect up to nine days' support in the first year, reducing in subsequent years to one day per term where there were no concerns and substantially more for UTCs facing significant challenges. UTCs preparing for Ofsted inspections also benefited from additional support, including training for governors.

Fourteen UTC principals left their posts between September 2015 and December 2016, a much higher number than expected. Some left for career reasons such as leading a larger school, others for family reasons and some because their appointments had not worked out well. BDT had no formal role in appointing principals, but was able to offer advice and support during the recruitment process. In addition, BDT worked with the principal of the JCB Academy, Jim Wade, to provide professional development workshops for new principals, especially those with limited senior experience. Joanne Harper, principal of UTC Reading, was asked to consider ways of meeting the professional development needs of more established principals, such as executive coaching and helping principals come together in groups of three to share ideas and experience.

A NEW SECRETARY OF STATE

Following the European Union membership referendum in June 2016, David Cameron resigned as Prime Minister and was replaced by Theresa May. Justine Greening took over as Secretary of State for Education. Nick Boles was replaced by Robert Halfon and Philip Hammond took over from George Osborne. Lord Baker bumped into the new Chancellor and his predecessor soon afterwards:

> [I] had lunch in Ebury Street. I saw the new Chancellor, Philip Hammond walk in and he was joined by George Osborne. After lunch I went over and briefly chatted to them. Immediately George said, 'I see you are wearing your UTC badge,' which I thought was a wonderful introduction. I replied 'I want to thank you George, for you were the biggest supporter of them'. Hammond said, 'I know all about them. I am a strong supporter – they are good news'.
> (Baker, K., 2016c)

In September, Theresa May announced plans to end the long-standing ban on new grammar schools. The idea had been trailed in the weeks before her speech at the Conservative party conference, created a great deal of controversy. Writing in *The Times*, Emma Duncan said:

> The British political class's obsession with selection prevents the country from focusing on the real problem with the education system, which is our failure to create a vocational strand ...
> University Technical Colleges (UTCs) – schools that provide vocational education for 14 to 19-year-olds ... need both money and official muscle.
> If Justine Greening ... wants to go down in history as the woman who set our education system on the right direction, it's the best thing she could do.
> (Duncan, 2016)

As Lord Baker noted in his diary, it turned out that the new Secretary of State was indeed sympathetic towards UTCs:

> To a meeting with the new Education Secretary, Justine Greening. I was told by her office that she would just squeeze me in between 1 and 1-30 in the House of Commons. So I prepared some papers about UTCs and when I got there it was just her, her Permanent Secretary, Jonathan Slater, and her private secretary. I explained to her the whole purpose of UTCs, how we have got to the stage of showing them to be successful, and could we please have a real commitment of support from the Department. The problems we have in the schools could be helped if only the DfE would do the right things. Very sympathetic, clearly liked it, and referred several times to the interview that I had done with Peter Hennessy about the importance of skills.
> (Baker, K., 2016d)

Justine Greening visited UTC Oxfordshire (Didcot) on 18 October and described it as inspiring:

> I think UTC Oxfordshire is fantastic for young people in and around Didcot, who have now got this brilliant new choice in terms of their education. I also think it's great for employers. The best scenario is one where employers get the skills they need for their future workforce and young people get the chance to build great careers. The UTC is an inspiring way to bring those two things together.
> (UTC Oxfordshire, 2016)

MEASURING STUDENT PROGRESS

Exam results showed some improvement in 2016, both at the end of key stage 4 and post-16. However, the government had introduced a new way of measuring performance at the end of key stage 4, known as Progress 8. The average Progress 8 score for all UTCs was minus 0.67 – significantly below the national average of zero. This was of course a matter of concern to BDT and DfE and there was undoubtedly pressure for UTC performance to improve.

However, that was not the whole story. For a start, Progress 8 was based on students' results in up to eight GCSEs and technical qualifications. Five of the eight had to be GCSEs in EBacc subjects – that is, English, maths, science, a foreign language, history and geography. Few UTC students took history, geography or a foreign language, and this directly reduced their UTCs' Progress 8 score.

Second, Progress 8 took students' achievements at the end of KS4 and compared them with results achieved in tests at the end of key stage 2 (age 11). In other words, Progress 8 measured progress over five years, but UTCs taught students for only two of those years. This mattered to UTCs because many of their students had underperformed in their previous schools. Indeed, some feeder schools actively encouraged students to move to a UTC if they were underperforming – for example, as a result of poor behaviour or persistent absence. Progress 8 was not designed to capture factors such as these. Peter Wylie (then BDT's Director of Education) pointed out to BDT's board that for UTCs to score well in Progress 8, they must make up any ground lost during the three years of key stage 3, but had less than two years in which to do so.

BDT argued that either UTCs should be exempt from Progress 8, or the measure should be adjusted to reflect the UTC curriculum. Initially, DfE did not agree with either proposition. Ofsted, however, was willing to consider other approaches to demonstrating the progress made by UTC students. With that in mind, a growing number of UTCs carried out baseline assessments of students when they joined in year 10 (age 14), which could then be used to measure progress over the following two years. Between September 2015 and February 2016, over 1100 students at 14 UTCs took CAT4 tests administered by GL Assessment, covering verbal, non-verbal, spatial and quantitative reasoning. By agreement with the UTCs concerned, the results were shared with BDT. In most UTCs, students' scores were below the national average. Some UTCs also used GL Assessment's reading age tests and English, maths and science knowledge tests.

Ofsted started to take account of UTCs' baseline testing results in their reports – for example in this 2016 report on the UTC in Sheffield city centre:

> [Sheffield UTC] data show pupils at the end of Year 11 made at least expected progress from their assessment on entry in Year 10 in English and mathematics. These data also show that disadvantaged students made as much progress as others in the college, from their starting points at the beginning of Year 10.
>
> Pupils with special educational needs made more rapid progress at Key Stage 4 than during Key Stage 3 at their previous school and the great majority of current pupils are meeting their targets.
>
> (Ofsted, 2016b)

In addition, Nigel Croft, head of Redborne Upper School and part-time member of BDT's field team, offered his school's help to provide a cost-effective data analysis service to UTCs. Participating UTCs received reports comparing their attainment and progress data with national averages and, over time, their own previous figures. This helped UTCs to set pupil targets and monitor progress, as well as providing aggregated data which could be used to highlight differences in performance (including differences between UTCs with similar technical specialisms) and to support conversations with Ofsted and the DfE.

In autumn 2016, BDT again collated data on student destinations. Out of 1292 18-year-old leavers whose destinations were reported to BDT by their UTCs, only 0.5 per cent were not in education, employment or training. Of the remaining 99.5 per cent, 44 per cent went on to higher education, 29 per cent started an apprenticeship, 15 per cent started a job, 9 per cent started other forms of education (e.g. at a further education college) and 2.5 per cent took a gap year. A highlight was provided by the Royal Navy, which offered 18 places for apprentice engineering watchkeepers: all but two were filled by UTC alumni.

ANOTHER CLOSURE

The UTC movement suffered a setback in February 2017 when it was announced that Greater Manchester UTC would close at the end of the academic year. *The Guardian* reported it in these terms:

> *£9m Greater Manchester college closes after three years due to lack of pupils*
> Not a single student achieved grade A*-C in both Maths and English GCSE last summer at technical school that has become seventh UTC to close
> (Weale, 2017)

Michael Gove was at that time a backbench MP and a regular contributor to *The Times*. On 10 February, he wrote: 'Some UTCs, like Reading, have been successes. But the majority have not. This week Greater Manchester UTC closed, just a few years after opening, the seventh to fold because of lack of numbers and poor performance.' (Gove, 2017)

Gove said the factors leading to UTC failure started with recruitment at 14; many parents, he said, felt that it was too soon to start on a 'narrowly specialist path' at the age of 14. He added that other schools saw UTCs as 'destinations for underperforming children', while higher-performing students were warned not to go to UTCs. He defended his belief in an academic key stage 4 curriculum:

> I wanted our schools to be more determined to give every pupil the means to become authors of their own life stories. And that principle lay behind my introduction of the English Baccalaureate ...
>
> UTCs perform very badly in EBacc rankings and consequently constrain students' future choices ...
>
> The lesson I've learnt about technical education is that what matters is developing high quality courses, not building shiny new institutions. But it's much easier for politicians to ask for, pose next to, and visit glossy buildings than ensure we have rigorous examinations in engineering draftsmanship.
> (Gove, 2017)

Professor Lord Bhattacharyya, a leading academic, industrialist, founder of the Warwick Manufacturing Group and trustee of the Baker Dearing Educational Trust, wrote to *The Times* about Michael Gove's opinion piece:

> The coverage of the closure of Greater Manchester UTC ... has rightly highlighted the importance of academic standards in all schools, but Michael Gove's dismissal of the UTC model shows he is out of touch ...
>
> UTCs equip students with the essential workplace skills and practical experience that puts them well ahead of their peers – and their destination results speak for themselves. This is why we recruit trainees, apprentices and degree apprentices from UTCs and will continue to do so.
> (Bhattacharyya, 2017)

Lord Baker's own response was published two days later. He, too, focused on student destinations:

> Michael Gove's attack on the quality of teaching and training in University Technical Colleges (UTCs) is ill informed. UTCs have the best employability rate in the country. There were 1,300 leavers at the age of 18 last July and only five became unemployed. A total of 44 per cent went to university (the national average is 38 per cent) and 29 per cent went to apprenticeships (national average 8 per cent); the rest obtained jobs or went to further education. The national unemployment rate for 18-year-olds is 11.5 per cent; for UTCs it is 0.5 per cent ...
>
> Every attempt to improve technical and hands-on learning since 1870 has failed – most killed by snobbery. UTCs will succeed because students want them, and the economy needs them.
> (Baker, K., 2017b)

RESEARCH

The National Foundation for Educational Research (NFER) published two reports on UTCs during 2017. The first, *University Technical Colleges: beneath the headlines*, was a contextual analysis examining the performance and characteristics of 37 UTCs that had been open for at least two years. Some features were already well known: for example, NFER found that almost two-thirds of UTCs were operating at below 50 per cent capacity, and average Progress 8 scores for UTC students were significantly lower than those achieved by peers in feeder scores. Less familiar was the analysis of student absence rates:

> At the start of Key Stage 3, absence rates for future UTC students are similar to their peers in the schools they attend during that phase. However, a significant difference emerges during the Key Stage 3 period, suggesting that there are some challenges with engagement for [future] UTC students during that period (Kettlewell et al, 2017, p. 4).

The report also suggested that headline accountability measures did not recognise the composition or breadth of curriculum offered by UTCs.

Recommendations included:

- independently assess UTC students at the point of entry
- review headline accountability measures
- review non-accredited technical and vocational qualifications offered by UTCs and if necessary, work with awarding organizations to develop suitable, accredited qualifications
- take action to address disincentives in the system which might hinder UTC recruitment
- commission research into higher-attaining UTCs

(Summarised from Kettlewell et al, 2017, p. 5)

NFER's second report was called *Evaluation of University Technical Colleges report – year one* (McCrone et al, 2017). This was the first stage of an evaluation of UTCs commissioned by the Edge Foundation and the Royal Academy of Engineering. The overall aim of the study was 'to understand effective practice and lessons that can be learned from the approaches currently being adopted, particularly in relation to curriculum design and employer engagement, as well as the broader challenges facing UTCs' (McCrone et al, 2017, p. 1).

Researchers found considerable employer awareness and

presence at all ten of the UTCs they visited. On the whole, UTC students 'were optimistic about the future, recognised that qualifications are important and that their UTC was providing them with the skills they need for the future' (McCrone et al, 2017, p. 2). They added:

> Interviewees believed that attending a UTC had benefitted young people in terms of: improved academic learning; enhanced technical skills and knowledge; increased transferable skills and readiness for the world of work; increased engagement with education and learning and motivation to succeed; greater awareness of and confidence in post-UTC pathways and increased likelihood of securing and maintaining chosen post-UTC destinations.
> (McCrone et al, 2017, p. 2)

Summing up the main challenges facing UTCs, the NFER research team said:

> The main challenges the case-study UTCs faced were ensuring that they secured and managed a suitable range of employers providing high-quality input into the curriculum; recruiting and retaining appropriate students with an interest in the specialism and who are motivated to engage and succeed; and recruiting and retaining high-calibre staff with appropriate knowledge and skills. Additionally, interviewees indicated that external curriculum changes, accountability frameworks and funding posed further challenges. Bearing in mind that UTCs are funded at the same rate as other schools, interviewees felt that the delivery of the UTC curriculum with associated costs such as more teaching hours, the purchase of specialist equipment and resources, and the cost of transport for work placements and workplace visits was demanding.
> (McCrone et al, 2017, p. 2)

17

THE JCB ACADEMY, PART 2

STUDENT NUMBERS, 2018-19

At the start of the 2018-19 academic year, 711 students were enrolled at the JCB Academy, compared with 575 the year before. The age range had been extended to create a year 9 intake for the first time. Applications substantially exceeded the number of places available, confirming the Academy's popularity and status in the region.

Having said that, relatively few key stage 4 students chose JCB Academy because of a clear ambition to become engineers. The chair of governors, David Bell, said:

> It would be lovely to think they all wanted to become engineers or business leaders, but that's not necessarily the case. Jim [Wade] carried out a survey, which selected every seventh child – so it was completely random – and asked, 'why did you come to the JCB Academy?' How many came because they wanted to be an engineer or a business leader? Out of 60 in the survey, only two.
>
> They might not have come here wanting to be engineers or business leaders, but the curriculum did influence their choice of coming to us. We've opened their eyes to a whole new world and they've actually got excited by it. And when you look at

> where they go when they leave here, you think – we've done a great job.
>
> (Bell, 2018)

Jim Wade commented on intakes at 14 and 16:

> I was a bit concerned before we opened that we would take predominantly from the bottom half of the ability range because of the image of vocational work in manufacturing. In fact, our [key stage 4] intake has been consistent since we opened: it's a narrow normal distribution, light at the top end and the bottom end, with a peak at the midpoint. For post-16 entry, we are more skewed towards the lower end. I think that's partly a factor of where we are, because all the schools round here have sixth forms.
>
> (Wade, 2018)

THE CURRICULUM

In 2018-19, key stage 4 students could take GCSEs in a range of subjects including:

- English literature
- English language
- Maths
- Chemistry
- Biology
- Physics
- Combined science
- German
- Spanish
- French

Other qualifications included:

- Foundation maths
- Financial education
- Business (BTEC First Award)

- Engineering manufacture (Cambridge Nationals)
- Engineering design (Cambridge Nationals)
- Principles in engineering and engineering business (Cambridge Nationals)
- Systems control in engineering (Cambridge Nationals)

In addition, all students took part in:

- Physical education
- Citizenship
- Enterprise education
- Careers education and guidance
- Religious education
- Personal, health and social education
- Work experience

Post-16 students joining the Academy's sixth form were offered technical pathways in engineering and business alongside A-levels in chemistry, physics, biology, maths, English literature, business studies and product design. Additionally, students could take an AS-level in quantitative reasoning and/or an extended project qualification.

Finally, students aged 16+ could join the Academy's apprenticeship programme, described in more detail in chapter 19, below.

The JCB Academy continued to emphasise project-based learning, delivered through a series of 'challenges'. David Bell was clear about the difference these made to student engagement:

> Why are kids engaged in our curriculum? Because they not just sitting in front of a whiteboard or a computer all day: they're doing stuff which is engaging and interesting and wherever we can, we apply that to their learning. (Bell, 2018)

Core challenge partners include Rolls-Royce, Bentley Motors, Harper Adams University, Rexroth Bosch Group, JCB (the business), Continental Engineering Services, National Grid, Toyota, the Royal Academy of Engineering and the Institution of Engineering and Technology.

Publicity materials emphasized the unusual nature of the academy's curriculum. For example:

> Our business diploma courses are quite different to the conventional way courses are delivered, offering significant enrichment to the learners who follow this vocational pathway. The opportunity to work with a variety of our challenge partner companies – Trentham Gardens, PWC and JCB to name but a few – provides our learners with authentic organisational issues to examine.
> (The JCB Academy 2019)

Key stage 4 engineering manufacture students took part in the Toyota Challenge, first developed before the Academy opened. Students started the project by looking at different methods of production and went on to manufacture a standard component in a variety of ways. As well as developing practical skills, students learnt about engineering production drawings, quality control systems and lean manufacturing principles. They visited one of Toyota's manufacturing plants to find out how Toyota implemented these systems and principles. Finally, Toyota set a challenge for students to solve, focusing on developing and applying lean manufacturing principles, teamwork and problem solving skills. They improved their communication skills by presenting solutions to Toyota staff.

In systems control, year 11 students worked on a challenge set by Network Rail and ULTRA-PMES. This started with electronic components: students learnt what they did and how they worked within a circuit. This led to a challenge based around an automated barrier crossing, in which students had to use CAD (computer-aided design) to design fault-finding systems in order to test and maintain the crossing control system.

Post-16 challenges included the Bosch Rexroth Challenge, which was tailored to a unit about hydraulics. Students visited a factory and conducted a series of practical exercises using Bosch Rexroth training rigs before visiting the London Eye to gain insights into the maintenance of hydraulic systems and sensors. In another challenge, engineering students spent two weeks at the Centre of Operational Excellence at the JCB Headquarters, focusing on lean manufacturing and quality improvement methods applied in JCB's assembly lines. There was a strong hands-on element, and students learnt about many of the tools and techniques used at JCB's manufacturing plants.

OFSTED

Ofsted carried out a short inspection in 2018 and awarded a grade 2 – good (Ofsted, 2018b). Inspectors noted that standards had continued to improve since the last inspection in 2014, in part because of effective use of pupil tracking data. Although a good proportion of key stage 4 pupils had been more or less disengaged from education at their previous schools, the Academy was 'highly successful in reintegrating these pupils back into their education and, as a result, they achieve good outcomes' (Ofsted, 2018b, p. 2). Behaviour was described as exemplary and pupils presented themselves as 'self-assured, mature, polite and friendly learners' (Ofsted, 2018b, p.2).

There was strong praise for the Academy's links with employers, too. The Academy's vision – shared with employers – 'to be a school with a strong reputation for educating the next generation of highly skilled engineers has become a reality' (Ofsted, 2018b, p. 2).

Ofsted found that post-16 standards were not yet at the same level as key stage 4, although post-16 outcomes were better in engineering and business than in other subjects. One factor was that a majority of students had until recently been following study programmes equivalent to more than three A-levels: following a review, the post-16 curriculum had recently been streamlined.

Jim Wade pointed out that some students left before completing post-16 programmes of study:

> A significant number of our year 12 students get an apprenticeship at some point during that year. Sometimes they come to us precisely because an apprenticeship is what they really want and their experience here gives them something to talk about at interview. Looking back to when we started, we hadn't factored that into our plans. If someone leaves

for an apprenticeship, you've got to see that as a success, but of course it has a negative impact on our performance measures – and our funding!

(Wade, 2018)

WHAT STUDENTS THINK ABOUT THE JCB ACADEMY

GROUP 1

The following comments were made during two group discussions at the JCB Academy on 23 January 2018. The first group consisted of six year 10 students (four male and two female).

Why did you choose to come here?

For the engineering aspect. I'm quite good at maths and I enjoy science so I thought it would be an ideal choice. My previous school tried to persuade me to stay. They said now I'd picked my options, I had to do my GCSEs with them.

Same for me, but also because of the sport we do on Wednesdays – I'm quite a sporty person. I'm from an army background, so I'm used to moving from one school to another.

Literally just the engineering. I really want to be an engineer. I'm a drag racer, so I've been fixing things for a long time – the hands-on stuff as well as the driving. I'd only been at my previous school for six weeks, so it wasn't an issue to move.

The stuff we do here will get you a better job in engineering.

My uncle works for the flying car company, so I know a bit about engineering.

At my old school we did design and technology, but it wasn't as in-depth as the engineering we do here. None of my friends came here from my old school. I knew only one person at the JCB Academy before I got here.

Did you find it easy to make new friends once you got here?

Yeah definitely. The Harper Adams trip made all the difference. Because we were working in groups we got to know each other a lot better. It helped make the friendship groups that we are in now. We did teambuilding activities and a project and a bit of learning – maths, for example, where we had to draw shapes. Instead of marks you got paid in fake money, so it was quite competitive. We decorated cups on the theme of Alice in Wonderland. The Harper Adams challenge involves building a remote-controlled car which has to survive going round a track with different bumps and gradients – the same track has been used since the first year group at JCB Academy.

Had you experienced working in teams in your previous schools?

In my old school, we had tables of five or six working together in most classes. Sometimes, we would swap groups.

That happened quite rarely at my old school. It was kind of individual education.

Same for me. It was very much, you're on your own, get on with it. We did sometimes get told to work in pairs and very occasionally in a group of four, but it was mainly do-it-yourself.

The groups we were in at Harper Adams carried on when we got back here – in engineering, that is. We had to carry on working together to design and build the car.

Engineering is definitely the subject with the most teamwork. I like the hands-on skills as well, the practical side. When working with the lathes, you work with a partner to get things done.

Once teachers get to know you, they let you help other students and the other way round. That obviously creates a good atmosphere.

The Harper Adams challenge was an introduction to the way the other engineering challenges would work and how they would be marked. We are now working on the JCB challenge, which is linked to our GCSEs. The JCB challenge is all about maintenance and the sustainability of materials used in vehicles.

We had a visit to the factory where they showed us a JCB 3CX [a tractor with a hydraulic scoop at the front] and described the features it comes with. Then they showed us how they service them. We had to write it all up on our laptops. We learned how the machines are made with maintenance in mind and about what makes them sustainable, how the materials are sourced and transported and how that affects the environment. It's about making a machine that doesn't damage the environment and can more or less last forever. They try not to use finite materials: things like steel can be recycled. We had to find out what different parts were made from, different plastics and all that – HDPE [high-density polyethylene], for example, which I had no clue about before! After we'd been told a certain amount, we had to go away and do our own research and write it up. We had to give examples to back up our own ideas and suggestions. The last part of the challenge is to do some maintenance on the engine: replace the oil filter, check there's nothing in the oil like rust or dirt, replace the pulley belt and check the alternator and water pump both work. That was interesting for me, because I'd literally never touched an engine in my life.

Do you have to make presentations, using PowerPoint and so on?

We had to do that for Harper Adams. Depending which group you are in you might present to your class or your engineering teacher. My group went to the Lodge [JCB Academy's apprenticeship facility] and presented to the head of apprenticeships. I felt more relaxed in front of him than I did in front of my classmates. I thought it didn't matter as much if I made mistakes, because I'd never see him again!

I definitely think this Academy makes you more confident and prepared to speak for yourself. That will definitely help when you go for a job. When you go for an interview you've got to be able to talk to the employer and this really builds your confidence.

We're currently working on a project where we're given a budget and told to build a model handler capable of moving several objects and withstanding damage when dropped from a certain height or carrying a certain weight. We made a first prototype before making the finished version.

Are subjects like maths and English taught differently here?

I'd say maths is quite similar, but they do point out how maths is used in engineering. My teacher was talking about using trigonometry the other day. The example he gave was the invention of the bouncing bomb for the dambusters raid and how they used lights on the aircraft wings to see how far up or down they needed to go before releasing the bomb. I thought that was interesting.

In English, there are differences. I find it hard to write long pieces, but here you get to use a laptop which makes it easier.

I'd say they're more understanding here about things like dyslexia.

Everyone does maths, English, science, engineering and business, and then you get one option. You can choose Spanish, French or German, so there's quite a wide variety of languages. There's product design, practical skills, sport and IT as well. We cover RE in form lessons, but we don't do history or geography. I would have liked to do history here, because I liked it at my previous school.

If someone was considering coming to JCB Academy, what would you say to persuade them?

There's a lot of demand for engineering skills and if you want a good career, this is the school for you.

For anyone who is dyslexic, there's a lot of support to help them express their ideas.

There's a big focus on science and maths as well as engineering, so if you wanted to do something else – be a vet or a doctor, for example – then this is a good place for you.

I wouldn't recommend it if you weren't at all interested in engineering, because we do at least an hour a day on engineering.

There's competition to come here and only a limited number of places. I'd have been crushed if I hadn't got it!

I think a lot of people don't understand engineering – how wide it is or how many jobs you can do. The technology side of music, planes, trains, vehicles, equipment – they all involve engineering. It's such a huge area. But I didn't know about it at my old school – they didn't talk about it.

I agree. You're not really told much about jobs in other schools. It's all about lessons – get your qualifications. But if you do know about jobs, you can really focus, find your skills and know what you want to do when you're older.

Work experience is important here, as well. Next year, I'll spend some time in my uncle's business and that will help me later when I apply for jobs or apprenticeships.

There's really good student support here. The door is always open if you need help or advice. There's a room you can go to if something's happened and you need to calm down and they are very good at managing bullying. There is absolutely zero tolerance of bullying, unlike my old school.

GROUP 2

The following comments were made during a group discussion at the JCB Academy on 23 January 2018. The group consisted of six year 13 students – four male and two female.

Why did you choose to come here?

I went to a conventional school and in sixth form I was studying business, geography and psychology. It all seemed very spread out and I didn't have a sense of direction. Coming to a more business- and engineering-driven school, I could see an end goal and was clearer about what I wanted to do. The different approach to learning here is outstanding – compared to my old school, definitely.

Four of us have been here since year 10. The change from a normal school to here was dramatic. The things we got to do in year 10 and year 11, we'd never have done in our old schools – going round JCB and Toyota, the things we've been able to use in the engineering department – we'd never have done those things, because no school is as well equipped as this one. Student support, as well: they care so much more here. We're like a big family.

I found out about JCB from a friend in the year above. I wasn't getting on very well at my old school – I was being bullied – and I needed to change school as quickly as I could. Coming here and seeing the facilities and how much support you get made me choose JCB rather than an ordinary school.

I came just for the sixth form, straight from GCSEs. I knew I wanted to do engineering – my dad's an engineer – and I didn't really want to stay at the school I was at, either. I applied for the Derby UTC as well as JCB, because it's nearer to where I live, but comparing the two places this one seemed better equipped and more established.

I first heard about this place when I was in middle school – about year eight – and talking to my friends about JCB Academy, they were all up for it. That got me thinking, let me have a look into it. Ever since then, I had a plan to get me to this point. From middle school, I went to high school for one year. There, I was kind of pushed against coming to JCB, but looking into it I found that the grades here, for what I wanted to achieve, were superior. The JCB name carries weight, too, and the smaller number of students here makes a difference – you're much closer to your teachers and tutors and you get a better relationship with students and teachers. Overall, the positives definitely outweighed the negatives. The dynamic and experience you get here are different from anything you would get in an ordinary school.

I came here because the subjects I could choose at my old school were very broad. I didn't want to choose geography as an option, for example, but when I found out about this place the pathways seemed more streamlined to engineering, which is what I wanted to do. You've got engineering design, manufacturing, various different pathways. Also, the facilities are very good here. We only had one workshop at my old school – here, we've probably got about six.

Did you already have in mind what you wanted to do?

Well, I've been open to inspiration throughout my school life. Coming here, I thought I wanted to go more on the business side.

I come from a farming background and I've always liked problem-solving and fixing machinery. From an early age I've always had that kind of mindset, so engineering appealed to me. I heard about plans for a Rolls-Royce Academy in Derby, but when it didn't open my mum looked around for other engineering-based academies and JCB came up. I've never looked back since coming here. The companies involved in engineering challenges, like JCB, Bosch, Toyota and Rolls-Royce, make all the difference. The Academy has a great reputation with different companies so if you want work experience or an apprenticeship, they assume we have the right attitude, which is great for all of us.

By the sound of it, you all have parents who supported your decision to come to JCB Academy. (Everyone nodded.)

Well, I did have to persuade them to let me come here, because it's a lot further than Derby UTC and I needed them to take me to the bus stop every day! But they were fine when I put everything across.

I had the motivation to do it, but it certainly helps to have the support of your parents.

How did your previous schools react?

We went to a parents' evening and said there was a possibility of me going to JCB Academy. Most of the teachers we spoke to said outright, we think you'd be better staying here. Looking back, I don't understand why that would be the case at all. This place is so different to any other school I've been to: why wouldn't you take up the opportunity to go somewhere like here? If you get the opportunity, you should take it with both hands. Normal high schools don't seem open to change. Maybe they don't want students to go because it has an effect on their grades? That's one of the reasons I came here, because on average they seem to be better here for what I wanted to do.

I think most ordinary schools think that you do GCSEs, stay on to do A-levels and then go to university. Round this table, there's three of us planning to do degree apprenticeships, all paid for, with jobs afterwards. That suits me, spot on. Had I stayed at my previous school, I would have struggled through A-levels, still not knowing what to do and would probably have drifted into university and accumulated debts of £50,000, and then I'd have had to find a job afterwards.

My school was more supportive. It's the only one in the area with a sixth form, so it's quite normal for students to leave and go somewhere else after GCSEs. What's more, other high schools were trying to feed their kids into our sixth form, so while the school didn't exactly encourage us to go, they did encourage us to look at other places before making our decision. I did have friends there – it's not like I had a horrible time – but it wasn't hard to move.

There was no way I could have stayed at my previous school and come out with a positive attitude at the end.

Can you tell me about the induction programme?

If I could go back to year 10 and re-do Harper Adams, I'd do it in a flash! I loved it. We went for a week. You know no one when you arrive, because you've only been at the school for two days. But when you get to Harper Adams, you're thrown in at the deep end.

You're in your houses, but you mix with everyone – all the different girls, all the lads – and do lots of activities every day. You couldn't get a better introduction week.

I think the organizers appreciate we all come from different places and they play on that very well.

I was the only one from my old school to come here, and apart from a lad in the year above, who I wasn't very close to, I knew no one. But pitched together at Harper Adams, you get to know people really quickly.

We didn't have that in the sixth form, though. We went to Alton Towers for a day, but it's not the same. And I couldn't do the teambuilding thing because I have a fear of heights! I made my friendship group simply by following people around.

Everyone is so friendly here. It's very difficult to find someone you wouldn't get on with. I was talking to a friend at another school the other day and she was saying it was very cliquey. You can literally sit down with anyone in the sixth form – everyone knows everyone else. It does really benefit the learning environment in the classroom, as well. It feels more chilled and more business-like. And teachers are not on your back the whole time.

Well, they'll chase you if you haven't handed some work in! I'd say it's like university: you take responsibility for your own work. Instead of being spoon-fed as you might be at an ordinary school, you're told what the deadline is and it's up to you to get the work done on time.

You start on that track in year 10 and you get used to it. By the time you come to sixth form, it's just an expectation that you will do the work.

What's been your experience of the way you learn here?

Some of the teachers are not really teachers – that is, they're from an engineering background or they worked in business before – which means they know what goes on in industry and in modern-day life. They can give you examples where engineering is used, which makes it easier to absorb information. Like in maths, if there's an equation, we can link it to what we are learning in engineering.

I especially remember that in GCSE maths – in English and maths, actually – even if it was only a short part of the lesson, teachers made a point of showing us connections with engineering.

It was nice to see everything linked together. I don't think you get that anywhere else.

Yeah, [in other schools] teachers just say 'this is what it is – just know it', not why.

You can't create a deep understanding of what you're learning that way.

Here, you're not just doing the course work, you're applying it.

Do you get to meet a lot of people from outside the academy?

Yes. For example, we went to the London Eye to learn how Bosch service the hydraulics. We got to see all the machinery down there and when we came back to the Academy there were sessions taught by professional engineers from Bosch. Having that relationship between students and actual engineers who know their stuff is really helpful.

There was that India trip, as well. Business students went out to see the JCB plant and we also went to the PWC offices to find out what they do over there. So even in India, they've got the links. We went with David Bell [chair of governors], who is obviously very high up in JCB and he showed us round and introduced us to people: it was good to see a business operates on a global basis instead of just over the road in Rocester! It opens your eyes, doesn't it? JCB's not just a local business, it's everywhere.

Meeting people from outside school certainly builds your confidence. My confidence certainly lacked something, especially when it came to presenting or even talking to new people. [Indicating a friend] you're the same: compared with when you started here, your confidence has improved loads. Getting to meet new people and talk to people who are higher up – people you'd never normally meet – it builds your confidence and makes you see that actually, when it comes to applying for jobs and stuff, you can actually talk to these people. You don't have to be scared or worried because they understand where you're coming from.

I don't think there'd be a chance I would have got the apprenticeship I have if I hadn't come here. I might have applied, but I wouldn't have got it. No way.

The Academy isn't just about the English and the maths and all that, it's about making you a well-rounded individual.

We've already touched on this, but what are you all planning to do next?

I've been offered a business degree apprenticeship at JCB, but I'm still comparing options and opportunities. I've been offered an assessment centre at Morrisons, and I'll go along and compare and contrast. I've also applied for university: I got five offers, two of them unconditional. That's my backup plan, but deferred for a year. That way, if I don't enjoy my apprenticeship I know I can still go to university if I really want to – but I can't see any reason why that would happen.

I'm joining the Navy to be a medic. I've always been interested in the forces and when I first came here I thought I'd go down the engineering path, but after work experience in the medical field I realize that actually, I still want to go into the forces, but with a different career route in mind. I've got through interviews and just have my medical and fitness tests to go.

I feel like a bit of an anomaly! I haven't applied for any apprenticeships because I'd like to take a different route. The English and maths teachers here are exceptional and I've been inspired by them. I know that's quite a cheesy thing to say, but I have! I'd like to go down the journalistic route, so I've applied for English at university. I've had four offers so far, and I'm waiting for one more response. That said, I am looking at apprenticeships as well because there are some being offered in journalism – but they are at level 3, which would be a kind of side step.

I've been offered in engineering degree apprenticeship at JCB. I've also applied to university and had four offers so far. For me, the degree apprenticeship offers the best route.

I've also been offered a degree apprenticeship with JCB, in engineering. I've also applied for uni and have five offers, three of them unconditional. But I don't really want to go to uni. Grade dependent, I'm definitely going to take the apprenticeship. It's too good an opportunity to miss!

I agree. But I can guarantee that if you sat down in a normal school, it would be university, university, university. Compared with friends at my old school, I can't name a single one who is doing an apprenticeship.

All these connections have helped us realize that there is a world outside school.

My little brother is doing his GCSEs this year and is applying for an apprenticeship. I don't think he would have done if I hadn't come here and looked at engineering. He'd have just gone with the flow and done A-levels instead.

It's a shame more people don't know about apprenticeships.

They don't really advertise or talk about apprenticeships in most schools in year 11, do they? If you say you're interested, they'll talk about it, but you have to bring it up in the first place. You're the one who has to say 'I'm going to do this'.

Is there anything you would change about JCB Academy?

Use the facilities more. In year 10 and 11, we got taught how to use the lathes and all that, but the opportunities are much more limited in the sixth form. But maybe that's about the courses I've chosen.

Personally, I can't think of any ways to improve the Academy. I'm happy here, and doing quite well. Because of the way we are supported, the way lessons are taught, I couldn't really fault it.

And we've all got clear destinations in mind.

Everyone is very helpful, whatever you want to do when you leave.

18

EMPLOYER ENGAGEMENT

Employers were central to Lord Baker's vision for UTCs. They would have a leading role in developing plans for new UTCs, be strongly represented on boards of governors, support curriculum projects, provide work experience placements, offer careers advice, provide apprenticeship places to former UTC students and offer tangible support ranging from bus subsidies to the donation of specialist equipment. In 2011, he said:

> The UTCs which we plan to announce in autumn 2011 have the support of 78 companies. This degree of commitment has the power to transform education. It's not the job of industry to build schools – that's the state's job – but industry can help enormously to equip schools and support the teaching that goes on inside. This level of support has never gone on at this scale in the past.
> (Baker, K., 2011)

In the early days, Lord Baker hosted or attended meetings and dinners with many employers across the country. BDT's patron, HRH The Duke of York, invited employers to events at Buckingham Palace. They came; they liked what they heard; they were willing to give time to support UTCs. They believed UTCs would provide new opportunities to attract young people generation into STEM careers, including pathways to apprenticeships.

Toyota was one of the earliest supporters of the UTC movement. At a seminar in 2015 a Toyota manager, John Byrne, explained why:

> Toyota wants manufacturing to be an aspirational career choice as it is in Germany and the United States. About 10 years ago, however, the company was struggling to recruit people of the required calibre, so we decided to develop a new apprenticeship programme. It was so successful that it was extended to the supply chain, especially small and medium sized businesses. Year by year, apprentice numbers grew from 20 or 30 to over 100, with 20 or more companies throughout the East Midlands taking part in any given year. It's not purely altruistic from Toyota's point of view: our suppliers have to have a skills plan which ensures they have the right people, with the right skills, to suppliers on time with what we need. Apprentices can progress to Higher National Certificates and Diplomas, full degrees and even Master's-level qualifications.
>
> For all that, the young people we were seeing were not altogether what we were hoping for. We therefore got involved in STEM showcases in both primary and secondary schools as well as further education colleges. We also encourage current apprentices to make visits to schools and colleges. Equally, schools bring students to visit the factory and to have a go at hands-on skills.
>
> We were approached by JCB to help set up the first UTC [the JCB Academy]. The challenge was to enhance the curriculum through good employer projects, linked to the requirements of the qualifications. This results in good experiential learning, but it is vital that the person presenting the project in class is from the world of work – that makes it entirely authentic in the minds of the students. We expect students to follow the same safety standards as our apprentices and behaviour has to be spot-on.
>
> We adopted the same working process as a lead sponsor of Derby Manufacturing UTC [opened in 2015], aligning projects with the curriculum and qualifications. As well as companies the size of Toyota

and Rolls-Royce, it's really important to involve smaller, local businesses. They benefit directly by meeting employees of the future and making young people and parents aware of the range of careers on offer in Derby. If they can spare just one hour in an eight week period, the UTC can make that work.

Of course, mainstream schools could do this, too. Unfortunately, too many see it as extracurricular activity which they can't afford to do because it cuts into core curriculum time: they don't realize the benefits it can provide to their students in bringing learning to life.
(Byrne, 2015)

JCB Academy students were introduced to Toyota's manufacturing principles, including 'kaizen' – literally, continuous improvement. They were expected to apply the kaizen approach to the 'Toyota Challenge', which involved designing and manufacturing fuel spacers for an engine. The winners of the first Toyota Challenge received a Duke of York Award for Outstanding Achievement. Toyota was also represented on the JCB Academy's board of governors.

Many of the UK's largest engineering, manufacturing, construction, media and IT companies became sponsors or supporters of UTCs, alongside large public sector organizations such as NHS Trusts. In addition, a great many small and medium-sized businesses were able to play their part. One early example was Haughton Design (see chapter 7, above); others included Morgan Tucker, civil and structural consultant engineers in Lincoln; Apps for Good in London; Flamingo Land in North Yorkshire; Compton Hospice in Wolverhampton; and Allcooper, security and fire safety systems specialists in Gloucester.

The armed forces also formed close links with UTCs. In 2015, the First Sea Lord, Admiral Sir George Zambellas, presented a charter to UTC Plymouth to symbolize relationship between the Royal Navy and the UTC. The Navy also played a leading role in establishing the UTC in Portsmouth.

Plymouth and Portsmouth have naval connections stretching back centuries. However, the Royal Navy saw an opportunity to raise awareness of careers for technically-minded young people across the whole country. As well as becoming a named supporter of several UTCs, the Navy invited all UTCs to enter competitions such as the Royal Navy UTC Young Engineers Challenge – Operation Amphibious Angel – in March 2017. In total, 212 students from 18 different UTCs took part in that challenge, which was based on the Royal Navy's experience of providing disaster relief. Working in teams, the students had to design a remote-controlled vehicle capable of operating on land and water. Quoted in an online article posted by the Royal Navy, Commodore Andy Cree said:

> The event is about inspiring the engineers and scientists of tomorrow and encouraging innovation. It enables the Royal Navy to highlight the variety of exciting roles and opportunities available to engineers, not just within the Royal Navy, but across the whole engineering and scientific community.
> (Royal Navy, 2017)

In publicity materials, BDT wrote:

> The Royal Navy welcomed over 80 UTC students and staff at their annual UTC Residential Visit Weeks held in early October. In a packed four day programme, students spent time at HMS Collingwood and HMS Sultan, the Royal Navy's Engineering Training Establishments, learning about the variety of roles for engineers in the Royal Navy. In addition to carrying out practical engineering tasks and completing leadership and team building activities, they also managed to squeeze in a visit to the Type 23 Frigate HMS Westminster whilst on a tour of Her Majesty's Naval Base, Portsmouth.
> (Baker Dearing Educational Trust, 2017c)

Interest in UTCs extended beyond the armed forces to the government's intelligence, cyber and security agency, GCHQ (Government Communications Headquarters), which was a founder sponsor of Scarborough UTC and later went on to welcome other UTC students to take part in courses and competitions aimed at inspiring young people to consider cyber security as a career. In a letter to Charles Parker, the head of GCHQ's station in Scarborough explained that:

> For a number of reasons GCHQ is struggling to recruit people with the right skills (STEM) in a competitive market ... By working closely with UTCs (piloting the idea in Scarborough) we would like to establish a pipeline of talented youngsters to help meet our recruitment needs ... [We] were so impressed by the potential we saw that we were able to go as far as to equip a Cyber Suite in the Scarborough UTC and to publicly avow our involvement.
> (Government Communications Headquarters, 2018)

The National Health Service, too, welcomed the opportunity to support UTCs. Soon after Liverpool Life Sciences UTC opened (see chapter 10, above), plans were confirmed for the Health Futures UTC in West Bromwich, sponsored by the University of Wolverhampton and The West Midlands Ambulance Service NHS Trust and supported by other health trusts, providers and pharmacists across the West Midlands.

Back in the commercial world, it was not only engineering businesses that saw the value of working with UTCs. Global Radio joined forces with the University of the Arts London to open The Global Academy UTC to focus on the technical skills needed in the broadcast media industry. Global's founder, Ashley Tabor, described his company as 'a young, dynamic and forward-looking business', but went on to tell BDT that:

> We currently struggle to recruit enough of the next generation of young talent who will help us achieve our goals. We consistently find that young people coming out of full-time education fail to possess the technical experience and understanding that we need in our industry ... the creation of the Global Academy University Technical College will go some way towards rectifying this situation. The students who study at the UTC will gain the technical and technological skills that will become a vital part of the fast-moving digital media industry over the coming years.
> (Tabor, 2013)

The Edge Foundation's research director, Olly Newton, summarised the importance of UTCs' links with employers:

> One of the things which distinguishes UTCs is that each one is sponsored by a local university and by local employers. Crucially, this means that UTCs have well-developed relationships with local employers which offer opportunities for work experience, visits to work places, employer-led projects and talks by practitioners.
>
> This sort of employer engagement not only exposes young people to a greater breadth of occupations than they might be familiar with, but helps them to develop a better understanding of the skills which employers need. Students at UTCs often have highly developed problem solving, teamwork and communication skills arising from the project-based and 'learning by doing' focused curricula.
>
> A good illustration of the value of careers advice, information and guidance is the experience of Ashley Freeman, an 18-year old student at Aston University Engineering Academy (AUEA). Ashley's dad works for the local council and his mum is a receptionist at a local business, but working on cars with his dad sparked a passion for engineering.
>
> Ashley opted to go to Aston University Engineering Academy post-16 because it gave him the opportunity to study for a BTEC diploma in engineering alongside A-levels in maths and physics, and get the hands-on experience he enjoys. He's also a dedicated Sea Cadet and planned to join the Royal Navy when he'd finished university. However, a talk from Captain Andy Cree of the Royal Navy Marine Engineering Advanced Apprenticeship scheme put him on to a different path. Ashley says:
>
> 'The plan was always to have a career in engineering and join the Navy, so going to university was a means to an end. The careers talk made me think again. Instead of spending three years at university at huge cost, I could start at a higher

level, have more managerial responsibility, use the engineering skills I already have and earn a salary.'
(Newton, 2016)

CASE STUDIES

LONDON DESIGN AND ENGINEERING UTC

The principal of London Design and Engineering UTC (LDEUTC), Geoffrey Fowler, explained how he worked with Thames Water on a project delivered in autumn term 2016:

> Thames Water came in and explained a scenario to all our staff. Teachers then had to work out how they would – individually – include the theme of 'water' into their lesson plans that term. It didn't have to be in every lesson for a whole term, just some. It also had to link to GCSE content. Just as importantly, it had to result in something that could be included in an exhibition at the end of term.
> Teachers had 100 minutes a week and a couple of drop-down days [days when the normal timetable was suspended] to work on this. Some teachers found it quite challenging at first because they weren't used to working this way, and by the end of term, some didn't have anything for the exhibition.
> That was two years ago. Since then we've helped teachers get up to speed ... What we've found is that teachers with industry experience or vocational expertise typically embrace the approach really quickly, but actually staff from conventional teaching backgrounds often see it as a breath of fresh air, too.
> (Fowler, 2018)

In addition, post-16 LDEUTC students taking the Extended Project Qualification accepted commissions from a range of the UTC's partners. Prompted by the charity Water Aid, one team devised a virtual reality environment depicting an Ethiopian village as its first fresh water stand pipe was installed. Another developed an idea for illuminating Waterloo Bridge as part of a 'Totally Thames' project. They built 3-D printers from kits before designing and making a model of Waterloo Bridge, which they equipped with colour-changing lights supplied by another industry partner. The colours could be changed remotely, simply by sending a tweet. In a third project, a robot was taught to ski.

RON DEARING UTC

At the UTC annual conference in 2018, the principal of Ron Dearing UTC in Hull, Sarah Pashley, outlined several curriculum projects being delivered in 2018 and 2019:

> Year 10: a project supported by Air Products, manufacturer of industrial gases: how computers are used to inform maintenance strategies, based on statistics. Students find out how Air Products use this technology to maintain and control multiple sites across the UK from their depot in Hull
> Year 11: introduction to Spencer Group's work with the rail industry, leading to a mechatronics project
> Year 12: Spencer Group, a multidisciplinary engineering business, commissioned students to generate a CAD design for the deck section of a ship's bridge, taking account of –
> 1. Structure and profile
> 2. Materials
> 3. Drainage
> 4. Wearing course (surface)
> 5. Services and housings (electricity, cabling, data links etc)
> 6. Section joints
>
> Year 12: students examined the mechanical fixing between a wind turbine blade and a rotor shaft, and explored the relationship between the blade manufacturer, Siemens Gamesa Renewable Energy, and wind farm operators

> Year 12: with support from RB (a consumer health and hygiene company), students worked on designs for control architecture and networks for a new plant
>
> Year 12/13: students worked on labelling machine maintenance project devised in partnership with Smith & Nephew, a multinational medical equipment manufacturing company
> (Pashley, 2018)

Pashley explained that curriculum projects formed only part of her UTC's approach to employer engagement. Employers also supported short projects, presentations and events linked to employability skills, project management, health and safety at work, and entrepreneurship.

At least once a term, employer representatives meet the principal and members of her senior team to review recent activities and discuss plans for the coming months. In September 2018, for example, they discussed work experience placements that had taken place over the summer. As the UTC only opened in 2017, this was the first time there had been a work placement programme. The timetable had been collapsed for a day to prepare students for their placements but even so there were some misunderstandings on both sides. One employer wrongly assumed that all students had learned to use CAD (computer aided design) software. Conversely, the same employer had not appreciated that some students were more involved in marketing and business development than design. As for students, most had been well prepared, but a few had not taken on board all the advice they had been given: for example, some had not completed a full written application and had therefore missed out on their preferred placements. That said, the programme was viewed as a great success and all employers were keen to support a longer (two day) workshop to prepare future students for work placements.

The employers were equally keen to support the UTC's student mentoring programme. Each mentor would meet a group of five students six times in the academic year to discuss their progress and provide advice and encouragement.

The UTC carried out a survey of 89 students starting year 13 in September 2018. By far the majority – 65 – said they hoped to find an apprenticeship place when they left the UTC, while 24 said direct entry to higher education would be their first preference. In practice, some would combine both routes by opting for a higher or degree apprenticeship.

WHAT DOES GOOD EMPLOYER ENGAGEMENT LOOK LIKE?

As with so many aspects of running a UTC, employer engagement involves two challenges: capitalizing on employer enthusiasm at the start, and maintaining momentum once the UTC is up and running. Principals typically took personal responsibility for liaising with employer supporters before and immediately after opening UTCs, subsequently delegating much of the day-to-day responsibility to curriculum leaders and individual members of the teaching staff. In BDT's view, there should always be a senior member of staff with explicit responsibility for maintaining effective links with employers, and in fact that is one of the criteria adopted by BDT to assess the strength and depth of UTCs' employer engagement programmes, alongside five others:

1. Employers see the UTC as key part of their talent pipeline
2. A senior level individual is responsible for employer engagement
3. Strong governance is driven by employers
4. Engagement is planned and "employer time" is ring-fenced
5. Processes and systems support effective engagement
6. Engagement is core to marketing, brand, strategy and messaging

(Baker Dearing Educational Trust, 2018)

When BDT staff reported to trustees in March 2018, they estimated that more than 730 employers were actively supporting UTCs. Their levels of engagement varied, but over half of were proactively involved in governance, projects, student recruitment and open events, or in some cases all of these activities.

In a briefing paper for the Edge Foundation's trustees (Mann and Virk, 2013), the Education and Employers Taskforce characterised most schools' engagement with employers as 'superficial'. In UTCs, by comparison, it was 'profound':

Looked at from an institutional perspective, the historic engagement of English schools with employers in general can be seen as superficial – not uncommon, but low volume and largely focused on "pupil progression" – that is, introducing pupils to the world of work. Only a minority of schools have routinely engaged employers in supporting teaching directly or through providing teaching materials or support to senior managers …

By extension, teaching staff (including senior leaders) rarely encounter employers and do not see them as natural partners in either teaching and learning or institutional operation. Employers do not sit alongside teachers as they design the curriculum; they are not regularly involved in delivering curriculum projects; and they are not automatically included on governing bodies.

Superficial engagement is typically –

- episodic
- non-iterative
- limited to narrow aims and purposes
- "bolted on" rather than "embedded": it is not part of the school's culture

The pattern of employer engagement found in UTCs is –

- broad – stretching across a wide range of activities and involving both staff and pupils
- deep – engaging individual employers in multiple activities relevant to young people through their school careers
- embedded – an accepted part of the UTC culture, regularly encountered by students and staff alike.

In the context of English education, this can be described as "profound" engagement. It stands in contrast to the relatively "superficial" levels of employer engagement encountered in most English secondary schools.

Profound engagement is further identifiable through three distinguishing characteristics:

- High volume. Staff and pupils within a typical UTC would be expected to engage with employers on many more occasions than peers across wider secondary education.
- Varied in character: Staff and pupils within a typical UTC would be expected to engage with employer across a much wider range of activities than peers across wider secondary education.
- Strategically integrated: Employer engagement within a UTC would be expected to sit firmly within coherent approaches to teaching and learning and pupil progression. It is part of the culture of the organization.
- Profound engagement can be seen as a process: through a structured arrangement, an individual employer or small number of employers provide pupils and staff with access to multiple contacts relevant to specific objectives. However, it is also an end in itself; it is itself an outcome of the school's culture.

(Mann and Virk, 2013, pp. 2-3)

This assessment was backed up by the first stage of an evaluation of UTCs commissioned by the Edge Foundation and the Royal Academy of Engineering. Researchers from the National Foundation for Educational Research (NFER) visited ten UTCs over a period of several months in 2017. The research indicated:

> … considerable employer awareness and presence at all the UTCs we visited. All ten case-study UTCs demonstrated moderate and contextual employer input into young people's learning. For example, through activities such as real life application of theoretical learning into the practical world of work; informing the curriculum with current industry skills' needs; observation and experience of every day industry activity; genuine, authentic challenges or problems for young people to solve; ongoing, regular input into projects; provision of visits to employers' workplaces; employer talks; resources and facilities; and specialist sector expertise.
> (McCrone et al, 2017, p. 2)

The second phase of NFER's evaluation looked in greater depth at three UTCs which demonstrated profound employer engagement, defined as 'where partners typically take ownership of a project; input into formative assessment; influence the delivery of curriculum components; and inform teaching and learning with specialist, current, technical skills and knowledge' (McCrone et al, 2019). Looking at the impact of employer engagement on students, the research team reported that:

> Students recognised that the projects and employer engagement benefitted their academic learning, technical and 'work ready' skills. They were aware that they were acquiring appropriate workplace behaviour, communication and interpersonal skills, developing their problem-solving skills as well as learning industry-relevant skills and knowledge. Additionally, UTC staff interviewees pointed out that young people's confidence had improved through working with employers, their understanding of the way the world of work operates had progressed, and their decision-making was considered to be better informed.
> (McCrone et al, 2019, p. 5)

19

APPRENTICESHIPS

From the very beginning, Lord Baker and Lord Dearing envisaged that UTCs would train apprentices. They hoped the Young Apprenticeship scheme would appeal to 14-16 year olds joining UTCs, and that full apprenticeships at level 2 (intermediate) and level 3 (advanced) would be available to post-16 learners.

In the event, Young Apprenticeships were abolished after the 2010 general election (see chapter 6, above). Nevertheless, it remained possible for UTCs to offer post-16 apprenticeships, and Aston University Engineering Academy (AUEA) agreed to run a pilot programme which would help other UTCs decide whether (and when) to launch their own programmes.

FIRST STEPS

AUEA

AUEA's pilot scheme, an advanced apprenticeship in engineering manufacture, was launched in September 2012 with funding from the Skills Funding Agency and support from the National Apprenticeship Service. AUEA's delivery partner was Semta Apprenticeship Service; together, they delivered training and development for apprentices employed by a single West Midlands employer. The apprenticeship opportunities were widely advertised throughout Birmingham and following shortlisting, candidates took part in a full day assessment conducted by Semta. Candidates were then interviewed by two senior members of the employer's staff.

There was considerable discussion about the delivery model. AUEA wanted a concentration of off-the-job training and development in the first year, with proportionately more time in the workplace in the second and third years. The employer, however, wanted delivery to be spread equally across each of the three years, with two-week blocks of off-job development at the beginning and end of each year and day release throughout the rest of the year; most of the time, therefore, apprentices were in the workplace for four days every week.

This delivery model proved very challenging in the first year, partly due to limited staff availability at AUEA. In addition, the employer sometimes required apprentices to come to work on days they were scheduled to be at AUEA, and it was hard for them to catch up on sessions they missed.

The pilot programme benefited from one-off financial support from the Skills Funding Agency: without it, the scheme would not have been financially viable.

THE JCB ACADEMY

With the JCB plant just a few hundred metres away, it is not surprising that the JCB Academy also started to think about apprenticeships at an early stage. Ideas were prompted further when students started to leave at the end of key stage 4, having been offered excellent apprenticeship opportunities with local employers.

The JCB Academy's first 47 apprentices were recruited in 2013; all of them were employed by the academy's sponsor, JCB. The academy later told other UTCs that:

> The process for securing apprenticeship funding from the SFA is time consuming and complex. You will need to allocate time and resource to this process. The JCB Academy set up a project team who worked for an 18 month period prior to the first apprentice programme being delivered: the JCB Academy had to meet the full cost of this 18-month planning and development period.
> (The JCB Academy, 2015)

BROADENING THE SCOPE

BDT was concerned that apprenticeship programmes could distract leadership teams from their core mission, especially in the first few years after opening. In BDT's view, UTCs should focus their energy on recruiting full-time students, providing good general and technical education, securing good exam results and helping students progress to apprenticeships, higher education or other destination.

In any case, funding was a real stumbling block for any UTC wishing to have a direct contract with the SFA. AUEA had received seed funding from the SFA, while the JCB Academy had committed a significant amount of its own funds to support the lengthy development process leading to the recruitment of their first apprentices in 2013. Other UTCs would also incur significant up-front costs if they decided to become apprenticeship providers in their own right. Furthermore, the SFA would only consider direct contracts with new providers if they could guarantee training valued at £500,000 or more per annum.

Against this background, the SFA and BDT strongly advised that any UTC wishing to dip a toe in the water should start by working as a subcontractor to an existing apprenticeship provider, with a view to signing a direct contract at a later date. Tresham College, joint sponsor of Silverstone UTC, hosted a briefing event in 2015 to explain their ideas for subcontracting some apprenticeship provision to interested UTCs, and a small number of UTCs proceeded to form partnerships with further education colleges or independent training providers.

UTC Reading, for example, launched an apprenticeship programme in 2015 in partnership with a local employer, Peter Brett Associates, and Activate Enterprise, a wholly-owned subsidiary of the UTC's sponsor, Activate Learning. The UTC's role was to deliver training and development leading to a BTEC Level 3 in Construction and the Built Environment, a level 3 technical qualification and in due course, EngTech status (see chapter 9, above), while Peter Brett Associates provided insights to the working life of an engineer.

Later in 2015, BDT asked Michael Schuhmacher – formerly with Rolls-Royce, and first chair of governors at Derby Manufacturing

UTC – to report on opportunities for UTCs to deliver apprenticeship training. One option was to establish a community interest company (CIC) which would manage a contract with the SFA on behalf of participating UTCs and provide apprenticeship training and development as sub-contractors to the CIC. This did not go ahead, mainly because of changes in the way apprenticeships were funded and delivered.

In his report for BDT, Schuhmacher (2015) identified six UTCs that already delivered apprenticeship training as sub-contractors; the JCB Academy was the only UTC that had a direct contract with the SFA at that time. A further 12 UTCs expressed interest in delivering apprenticeship services. He also found that most UTCs lacked a full understanding of apprenticeship frameworks, funding mechanisms or delivery rules, and few had discussed apprenticeships with local employers. Schuhmacher recommended a programme of support to improve UTCs' understanding of apprenticeships, develop their links with employers and potential partners such as group training associations, and help them develop appropriate governance arrangements.

Meanwhile, the government pressed ahead with substantial reforms to the apprenticeship programme. Block contracts between the SFA and training providers would become a thing of the past: instead, employers would take the lead not only in appointing apprentices, but in contracting for the provision of relevant training and development. Providers selected by an employer would receive funding for the training they provided to individual apprentices. At the same time, apprenticeship 'frameworks' were to be replaced by new 'standards', and new assessments would be introduced at the end of every apprenticeship.

Faced with this level of complexity and uncertainty, most UTCs decided not to develop an apprenticeship offer – at least, not until the dust had settled on the apprenticeship reform programme.

EXPANDING APPRENTICESHIPS AT THE JCB ACADEMY

The JCB Academy proved to be an exception to the general rule and continued to expand its apprenticeship programme. Jim Bailey, who originally joined JCB Academy as a teacher, was appointed Director of Apprenticeships. Asked how he became involved, he said:

> I volunteered! I thought, that's a really good challenge for me. It was a risk because as a schoolteacher, I taught A-levels, so apprenticeships were a very steep learning curve. What I did have was a bit of industrial knowledge – I trained in product design and worked in industry for two or three years before retraining as a teacher.
>
> (Bailey, 2018)

By 2015, the Academy offered intermediate and advanced apprenticeships in engineering, fabrication and welding, business skills and information technology, not just for JCB (the business), but also for a growing number of other employers including Michelin, Alpha Manufacturing, Bri-Stor Systems, Orbital Gas Systems and Continental Engineering Services. By now, apprentices were travelling from Stoke-on-Trent, Staffordshire, Derbyshire and Nottinghamshire.

As numbers grew, it was increasingly difficult to accommodate all apprenticeship provision on the main Academy site. The former Dove First School, a short distance from the main building, became the Dove Engineering Centre. Facilities included machine, welding and electrical workshops, mechatronic and metrology equipment and a self-contained canteen.

The training model provided for first-year engineering apprentices to go to the Dove Centre for four days a week and to their employer's workplace for one day a week, throughout the first 12 months of their programmes. In the second and third years, apprentices spent most of the week at their workplace, but continued to attend the engineering centre part-time for off-job training and development. Links with Sheffield Hallam University meant apprentices could progress to a full engineering degree later. Jim Bailey said:

> We want technicians to be called by that name – craft workers likewise, because they have genuine craft skills and abilities. Engineers are notionally at the top [of the professional ladder], but they can't exist without all these skilled, quality people lower down. We want young people to recognise that there is a path to the status of engineer. We want them to find

the right first level, perhaps as a technician, perhaps as a craft worker, start a good career, add value for the employer and move up, perhaps eventually to degree level and beyond.

For that reason, all our engineering apprentices go through the same introduction at a grassroots level whether they're going to be an engineer with a degree or whether they're going to get a craft role. They're all going to be able to weld, be able to work with sheet metal, work with electrics, do some milling, fitting, drafting on a drawing board, CAD, metrology, maths. Whether they like it or not, for the first twelve months they all get two and a half hours a week of maths so that they know trigonometry, about indices, they can apply that maths correctly … and I consider that to be very valuable, one of the most valuable things we do.

The amount of time we take to deliver a unit would be the same if we spread it over twelve months: it's always 80 hours per unit, but instead of doing two hours a week over 40 weeks, we're doing 80 hours over three weeks. But because there's no setting up at the start of a session, no packing up at the end, no going over learning that we did last week and have already forgotten, learning is more focused. The learners tell us – because we ask them – they are more focused and they really enjoy this approach. We get very few behavioural issues because they're all on the case.

(Bailey, 2018)

APPRENTICESHIP REFORMS

UTCs seemed well placed to deliver apprenticeship training and development. They were sponsored and supported by a wide range of employers. They had industry-standard facilities and equipment, specialist teachers and tutors with relevant professional and industrial experience, and many full-time UTC students chose apprenticeships over other options when they leave at 16 or 18.

However, apprenticeships were in a state of flux when UTCs were being established. The introduction of a levy payable by larger employers involved a move away from block payments to training providers to a system based entirely on employer choice. In addition, frameworks were being phased out in favour of standards. The government wanted to encourage employers, colleges and training providers to switch from frameworks to standards at the earliest opportunity.

One way of doing this was to reduce funding for training linked to the old frameworks. In the case of engineering operatives, the new standard – once approved – was expected to attract total funding of up to £12000 per apprentice, while funding for the pre-existing framework would be halved. Training providers had a clear financial incentive to move over to the new standard as soon as possible.

The snag lay in the two words 'once approved'. The new standard was developed by a panel of leading employers, including representatives of JCB (the business), and presented to the Institute for Apprentices for approval. It was rejected several times over a period of many months, before finally being approved on 24 September 2018.

In the meantime, the government had already cut the funding rate for the old apprenticeship framework, leaving many colleges and training providers with a substantial, unplanned cut in funding.

The effect on JCB Academy was profound. The Academy had recruited 90 engineering operative apprentices, expecting to receive £12000 per apprentice. Actual funding amounted to only around a third of this figure. The Education and Skills Funding Agency provided a loan of £650,000 to cover the shortfall, to be repaid over a period of years. The episode was a sobering example of the risks involved in apprenticeship provision during a period of major reform.

Weighing up priorities and risks, most UTCs chose not to deliver apprenticeship training at all. A minority acted as sub-contractors to other colleges and training providers, and only the JCB Academy became a significant apprenticeship provider in its own right.

LEILA'S STORY

Leila Worsey was at JCB Academy from 2009 to 2011 as part of the first key stage 4 intake. She was 14 when she started and left at 16 to become an apprentice at JCB (the company). She said:

We had quite a lot of contacts with employers at JCB Academy, which was both exciting and in a way, reassuring, because we got an insight into what companies did and what to expect in working life. My ambition was to get an apprenticeship, and we were able to ask questions about what employers were looking for and what it's like to work at different places. That's what made me want to work for JCB after leaving the Academy – I really like the way they treat their employees.

I started at JCB as an advanced apprentice [level 3] at the age of 16. My job was in electrical engineering and maintenance. I didn't find it intimidating to be a woman in a mainly male workplace, which in hindsight I find a bit surprising, actually! There I was, 16 years old, the only female apprentice in my group, working on the shop floor with all these men – but I didn't find it daunting.

JCB is working hard to increase the proportion of women it employs. I'm a STEM ambassador and represent JCB at events for 12 to 16-year-old girls in secondary schools. I always ask them about their perception of engineering and sure enough, we get a lot of comments about car mechanics and lorry drivers. So then we explain what we do, we talk about seven different types of engineering, JCB Academy, apprenticeships and university courses. They start asking 'is this engineering? How about this?', and gradually they start to grasp what engineering really is. By the end of the workshop, some of them are very interested. Not all of them, of course, but that's fine – at least they've got a better understanding of what engineering really is.

And it's working. Over the six or seven years I've been there, I've seen a big intake of female apprentices including many from JCB Academy.

JCB gave me the opportunity to progress onto a higher apprenticeship in manufacturing integrated engineering, which I started in 2014. I completed a Foundation Degree at Sheffield Hallam University as part of that.

I'm 23 now and earlier this year I started a new role as one of seven quality engineers, helping to oversee the quality assurance and inspection teams who look at every product on the shop floor, recording any defects and feeding that back to the production team.

I had my mid-year review a couple of months ago and discussed my future with my line manager. He said the 5 to 10 year plan is for me to take over his role, which was really great!

I'm sure I wouldn't have got this far, so soon, without attending the JCB Academy.

(Worsey, 2019)

20
LOOKING TO THE LONG-TERM

On 9 January 2017, the recently-appointed Secretary of State for Education, Justine Greening, invited Lord Baker and Charles Parker to discuss the challenges facing the UTC movement. The Parliamentary Under-Secretary of State for the School System, Lord Nash, was also at the meeting, together with senior officials.

The Secretary of State had been impressed by visits to UTCs in Oxfordshire and Scarborough. *Scarborough News* reported her visit in these terms:

> *Justine Greening: Scarborough UTC is just what young people need*
> Ms Greening told the *Scarborough News* that today, she believes technical and hands on education is as essential as a university degree.
>
> She said: "I think it's fantastic for giving these young people a very different education - much more practical and much more technical and really one that they find incredibly stimulating [and] it's matching them up with amazing local employers that can offer them great opportunities and great careers."
> (Paterson, 2017)

Greening appreciated UTCs' potential and wanted them to contribute to a wider strategy for technical education. On the other hand, she understood that student recruitment had been weak and many UTCs were facing financial difficulties. If the movement was to survive and prosper, solutions would need to be found. She asked officials to conduct an internal review, but also took immediate action on three fronts.

First, local authorities would be required to write to parents and carers of all year 9 students about post-14 choices. This would improve awareness of UTCs, studio schools, colleges and other providers offering pathways for young people aged 14+. Second, UTCs would receive additional funding for a limited period. Third, there would be no further UTC application rounds for the time being.

Although delighted to have the Secretary of State's broad support (not to mention additional funding), Lord Baker believed letters to parents and carers would be only a partial solution to the recruitment challenge. He believed UTCs should have a right to speak directly to young people in their current schools. The Department for Education said this would require primary legislation.

THE BAKER CLAUSE

Lord Baker quickly identified a way of changing the law. Parliament was considering a Technical and Further Education Bill. The bill had been introduced primarily to give the Institute for Apprenticeships responsibility for college-based technical education as well as apprenticeships – specifically, to oversee the development and implementation of new T-level qualifications. The bill also introduced a new insolvency regime for further education and sixth form colleges, together with provisions on data collection and Ofsted's remit.

Lord Baker sought advice from clerks in the House of Lords. Would it, he asked, be possible to add a clause to the bill, requiring schools to allow UTCs – and perhaps other providers as well – to speak directly to their students? The answer was yes.

Lord Baker recalled a key moment:

> Just before I spoke in the Technical and Further Education Bill debate, [Lord] John Nash tapped me on the shoulder and said, 'the secretary of state has agreed

> to accept your amendment'. This is what we have been waiting for. He also said we can't announce it yet.
> (Baker, K., 2017)

During the first Lords debate on the Technical and Further Education Bill, Lord Baker said:

> I intend to move one amendment which will improve the Bill enormously, in my view, dealing with career advice. How are you to get knowledge of apprenticeships over to youngsters? You cannot expect the schools to tell them, because teachers leave their schools, go to teacher training colleges and then straight into teaching. They had no experience of government, industry and commerce, or of apprenticeships. I will move an amendment which will allow the providers of apprenticeships, along with the heads of university technical colleges, studio schools and FE colleges, to go into schools at 13, 16 and 18 to explain to the students what they can then study – the alternative offerings.
>
> I am glad to say that the amendment has the support of the noble Lord, Lord Adonis, the noble Baroness, Lady Morris, and the noble Lord, Lord Storey, from the Liberal Democrats, as well as of several Conservative Members, so I expect it to pass.
> (HL Deb 1 February 2017, col. 1220)

The essence of the Baker clause – as it became known – was that publicly-funded schools in England 'must ensure that there is an opportunity for a range of education and training providers to access registered pupils during the relevant phase of their education for the purpose of informing them about approved technical education qualifications or apprenticeships' (Technical and Further Education Act 2017, section 2). The 'relevant phase' started in the year in which the majority of pupils attained the age of 13 – that is, year 9 – and finished at the end of the year in which a majority reached the age of 18.

The new clause was debated and approved by the House of Lords Grand Committee on 22 February. Lord Baker introduced the clause in these terms:

> The curse of our education system at the moment is that secondary schools or comprehensives seem to have only one target: three A-levels and university ... There are many pathways to success and it is our duty to try and open them to more people.
>
> The amendment will strengthen the Bill significantly by giving all young people the chance to hear directly from providers of apprenticeships and technical qualifications about what they can study ...
>
> Of course, having [a] transition at 14 presents marketing difficulties because youngsters, having gone to an 11-16 or 11-18 school, do not expect to make another choice until they take GCSEs. Certainly, UTCs have had difficulty recruiting at 14 ... many schools resist anybody who comes in and tries to persuade a pupil to go on another course. It is a loss of money – about £5000 a head – and they are very hostile ...
>
> Sometimes we have to take difficult students, and I am very proud of the fact that we have remarkable examples of turnarounds where a student's life opportunities have been fundamentally changed. One has to recognise that the famous key stage 3 for 13 and 14 year olds is a very troubled stage indeed. You have a large number of disengaged and uninterested students and that is not getting better; it is always there. We are providing an opportunity for them, and we also provide a great opportunity for talented students.
>
> (HL Deb 22, February 2017, cols. 53-70GC)

The Liberal Democrats' leader in the House of Lords, Lord Storey, supported the amendment. 'We have this tramline approach in this country that there is only one route to go down. We need to break free of that,' he said (HL Deb 22, February 2017, cols. 53-70GC).

The Labour peer, Baroness Morris of Yardley, supported not just the amendment, but UTCs as well. 'I agree with Lord Baker that the UTCs are a force for good. They had a difficult birth and baptism but they are still a major player in the field. In a way, they encapsulate the problems of the incentives in the system. Their very existence is threatened because we have the wrong incentives' (House of Lords GC debate, 22 February 2017, column 60).

Responding to the debate, Lord Nash thanked Lord Baker for tabling the amendment and paid tribute to his work in establishing the UTC programme – 'I particularly enjoyed his unbiased commercial for them'(HL Deb 22, February 2017, cols. 53-70GC) – before accepting the amendment on behalf of the government. He said:

> Schools will be required by law to collaborate with UTCs, studio schools, further education colleges and other training provides. This will ensure that young people hear more consistently about alternatives to academic routes and are aware of all the routes to higher skills and into the workplace. This is vital if we are to set our technical education on a par with the best in the world. I thank my noble friend for this thoughtful amendment and I accept it.
> (HL Deb 22, February 2017, cols. 53-70GC)

The Baker clause came into effect on 2 January 2018. It was welcomed by the further education sector and of course by UTCs. The schools sector, however, was less enthusiastic: indeed, very few schools immediately acted on it.

MINISTERIAL DECISIONS

On 4 March, Justine Greening wrote to Lord Baker to confirm decisions taken after their meeting in January. She started with a succinct statement of the problems to be solved:

> While the destination data that Baker Dearing collects at age 18 appear to be positive, exam results at 16 suggest that UTCs are not working for enough of the young people who attend them. The low pupil numbers and financial position of many UTCs suggests they are not working for the taxpayer…
>
> I want to repeat my commitment to the UTC concept. Unless we improve technical education in this country, we will not have an education system that

works for everyone. That is why I want to work with you to make sure that UTCs succeed.
(Greening, 2017)

The Secretary of State confirmed that transitional funding would be provided to UTCs at the annual rate of £200,000 per UTC for three years. This would be backdated for those UTCs already open for more than a year. In return, all UTCs would form partnerships with other schools and in some cases, join a strong multi-academy trust.

DfE would also provide funding for all UTCs in their pre-opening phase, so they could employ a principal for five terms before they opened instead of just two terms. This would allow a longer period of planning and preparation.

Greening confirmed that local authorities were expected to write to the parents and carers of year 9 students before 14 March, setting out choices at 14+. She noted that the Baker clause had been added to the Technical and Further Education Bill, and hoped this would have a positive impact on student numbers in years to come.

Baker Dearing had sought permission to set up a nationwide multi-academy trust for UTCs, the UTC Network Trust. The department believed UTCs needed strong partnerships with good and outstanding schools in their own local areas. The UTC Network Trust would not achieve this, and the department therefore did not give it the green light.

SKEWED RECRUITMENT

Principals continued to tell BDT that other schools encouraged challenging and low-attaining students to move to their UTCs. BDT trustees heard that:

> Some UTCs have been obliged to take in large numbers of students with learning and behaviour problems who are not well suited to being in a highly specialist school. There are numerous documented examples of students and their parents being "directed" to UTCs by schools and local councils as an alternative to permanent exclusion, with some schools putting this

advice in writing to the parents. Indeed one school … has twice sent its 45 worst performing students to [a] UTC. Because the Fair Access Protocol does not apply to direct admissions at 14 it has been hard for UTCs to resist these admissions; and even when oversubscribed if the student lives near enough to the UTC, they can insist on entry.

We have been seeking action about this issue for several years. There are two possible ways forward: to change the Admissions Code so that the Fair Access Protocol applies to admissions at 14; or to permit UTCs to make admission dependent on demonstrating aptitude for and interest in the specialism through a change to their Funding Agreement allowing them to dis-apply elements of the Admissions Code.
(Baker Dearing Educational Trust, 2017e)

In the summer of 2017, DfE officials reviewed the evidence at ten UTCs and found that some schools did indeed encourage students with a poor record to move to neighbouring UTCs. BDT lobbied for admissions policies similar to those used by City Technology Colleges, which would allow UTCs to consider applicants' motivation and aptitude for the technical curriculum. BDT sought counsel's opinion, which confirmed that such an approach would be lawful. The government did not agree to change the admissions code.

Increasingly, UTCs asked students to take baseline tests at the start of year 10 (age 14). Test results showed that many students had made less than expected progress during key stage 3 – in some cases, *much* less. Ofsted took this into account in inspection reports from 2017 onwards, as these examples show:

Senior leaders [at Lincoln UTC] have rightly recognised that a significant proportion of pupils arrive at the school in Year 10 behind in their communication skills, particularly in their reading skills. They have put into place appropriate strategies to resolve this issue. (Ofsted, 2017, p.3)

A good proportion of pupils come to [the JCB Academy] with behaviour and/or engagement issues, having been previously disengaged from their education. You are highly successful in reintegrating these pupils back into their education and, as a result, they achieve good outcomes and complete the phase of their education. (Ofsted, 2018b, p.2)

Staff [at Bristol Technology and Engineering Academy] provide a nurturing environment for a significant number of pupils who have not thrived in their previous schools. As a result, most pupils make better progress than they did before they joined the school. (Ofsted, 2018c, p.1)

Some pupils [at London Design and Engineering UTC] start the school having struggled to engage with their education previously. A review of case studies shows that the school has been successful in re-engaging pupils with learning and improving their behaviour, attendance and confidence. (Ofsted, 2018d, p.6)

Many pupils arrive at [Greater Peterborough] UTC having had an unsuccessful key stage 3 experience in other schools. This is particularly evident in their reading ability and literacy skills. (Ofsted, 2019a, p.1)

Leaders' own information shows that many [Bolton UTC] pupils arrive in Year 10 having made little progress in key stage 3. Inspection evidence shows that pupils currently in the college make strong progress. (Ofsted, 2019b, p.7)

Pupils' levels of literacy on entry to [Leeds UTC] in Year 10 are below average. Leaders ensure that the weakest readers receive frequent reading support when they are in Year 11. The school's own information demonstrates that pupils who receive additional support with their reading when in Year 11 make strong progress. (Ofsted, 2019c, p.5)

OPERATIONAL AND STRATEGIC SUPPORT FOR UTCS

By 2017, BDT regarded about half of all open UTCs as secure: that is, they were making good progress and managing challenges well. On the other hand, there were concerns about 25 UTCs, mostly as a result of low student numbers, financial pressures or weak Ofsted reports. Under Peter Wylie's leadership as director of education, BDT's education advisors maintained links with all UTCs, while focusing most of their time on UTCs in most need of help. BDT's trustees were reminded that in addition to the Director of Education:

Baker Dearing has five part time education advisors [who] are commissioned for either six or eight days per month, mostly in term time. Each of them has a group of UTCs for whom they are the initial link as well as providing support for UTCs collectively.

The team works alongside projects both in the pre-bidding and, once approved, the pre-opening phase. Their aim is to help the sponsors, who are often employers, get to grips with the requirements of opening and running a school. This can involve substantial time reviewing and editing education and finance plans, preparing curriculum and staffing plans or at least ensuring these are as robust as possible, and assisting with the preparations for "readiness to open" meetings and the Ofsted pre-registration inspection. The education advisor to the project then becomes the advisor to the open UTC to ensure as much continuity as possible.

Each open UTC receives at least one half day visit per term. This is designed to keep in touch with the Principal and to offer any support. We look at any issues the Principal thinks need attention as well as looking at recruitment, attainment and progress figures and any reported issues about safeguarding. It is important to recognise that UTCs are intended to be self-supporting though our team have worked hard for example on brokering relationships with Teaching School Alliances to make good use of the DfE UTC [transitional] grant. UTCs in their first year get at least two visits per term and more if required.

Our role is not to inspect UTCs, though of course we do not ignore things that need attention. These are stand-alone schools and the extent to which they accept advice and use our expertise varies. We focus our time and attention on those most in need of support. We are exploring ways we might increase our visibility and contact with governing bodies but in a way that represents a cost effective use of limited time.

The team has been closely involved in the discussions in UTCs about joining Multi Academy Trusts (MATs) helping governors work out the criteria they should be

using to evaluate MAT proposals and advising Principals on how to achieve an effective and fair process. We have also devised a partnership framework for UTCs to use where there is no obvious MAT to join.
(Baker Dearing Educational Trust, 2017b)

BDT worked closely with colleagues in the Department for Education and the Education and Skills Funding Agency (ESFA). In 2017, BDT and the ESFA ran three seminars for UTC finance directors and principals. Simon Connell, by now BDT's development director, helped several UTCs redraft their financial plans and brokered repayment schedules covering sums owing to the ESFA.

BDT also arranged briefings for UTC governors. At the annual UTC conference in July, Charles Parker highlighted a key issue: governors nominated by employers often had limited understanding of the way schools were run and held to account, while many teachers and senior staff in UTCs had limited experience of industry. Bridging the gap required hard work and commitment on both sides.

Charles Parker listed some of the lessons learned from governors' meetings across the UTC network:

- Student recruitment should be a top priority at every meeting
- Health and safety should be on every board meeting agenda
- Financial problems should be anticipated early
- Charity law applied: the governor's first duty is to the beneficiaries (the students) and to safeguard the assets (land, buildings, UTC brand)
- Remember to follow due process at all times
- Make sure you have a competent and genuinely independent clerk
- Don't be frightened of the principal – he/she works for you and needs your constructive engagement/support

(Parker, 2017)

BDT followed up by arranging a series of regional workshops for UTC governors. Peter Wylie spoke about the Ofsted inspection framework, highlighting the kinds of questions governors should ask to satisfy themselves that UTCs were providing a good education in a safe environment. Most were no different from the questions asked by governors in any secondary school.

For example, governors should understand student progress, attendance, achievement and so on, and satisfy themselves that their UTCs were upholding high standards in areas such as safeguarding and additional educational needs.

At the same time, however, governors had to consider the factors which made UTCs different from mainstream secondary schools – a point emphasised in a guide to UTC governance prepared by Simon Connell. The guide included key points to help governors reflect on their UTC's strengths and weaknesses, including:

- Students want technical specialisms, hands-on learning, state-of-the-art equipment and opportunities for work experience. Is your UTC offering this?
- Termly visits to a UTC during the school day is a vital part of a governor's on-going education and oversight role. When did you last visit your UTC?
- Student recruitment should be a top priority at each governors' meeting. Are you investing sufficient time and money to attract the right students?
- UTCs run very tight budgets. Indicators such as the pupil/teacher ratio and the percentage spent on teaching staff are good guides to gauging financial sustainability. You should know these measures for your UTC.

(Baker Dearing Educational Trust, 2017d)

MARKETING

During 2017, BDT stepped up its involvement in marketing and public relations, led by Anna Pedroza, Director of Marketing. In addition to a marketing conference open to all UTCs, webinars helped UTCs develop effective marketing strategies, engage potential students and implement consistent, long-term communication plans. About a week after each webinar, groups of five or six UTC principals could join a conference call to discuss actions, share ideas and seek further advice. In some cases, BDT provided one-to-one marketing support.

At a national level, BDT commissioned a research company, Emsi, to analyse labour market data to provide information on STEM occupations and industries and project trends to 2022. In addition, OnePoll surveyed a panel of 1000 adults aged between 20 and 35 and currently working in a STEM sector. Their questions covered perceptions of school experience, favourite subjects, the value of subjects to future careers, work readiness and the experience of applying for jobs. The results of these two pieces of work were set out in a report called *From school work to real work: how education fails students in the real world*, which BDT published in February 2017 (Baker Dearing Educational Trust 2017a). The launch event was attended by employers and professional bodies including Hewlett Packard, Atkins, Fujitsu and the Institution of Engineering and Technology.

The OnePoll survey pointed to a considerable gap between school and the world of work. Only about 40 per cent of respondents said their schools had offered insights into the labour market, and over half (55 per cent) reported not understanding the connection between the subjects they studied at school and their use in the world of work. A majority (61 per cent) felt that applied technical skills could have prepared them better for their careers than the more academically-focused education they received and a similar proportion (63 per cent) believed employers should have a greater input into what schools teach.

BDT's report said:

> The young STEM workers who took part in the survey were often students who enjoyed school but, looking back, recognised that it had not effectively prepared them for their career in STEM. The majority of young STEM workers surveyed found school too remote from the realities of working life. Many recall feeling disengaged because of the lack of connection between key skills like maths, English and computer science and their very real applications in the workplace. Two-thirds feel that they made choices about subjects without fully understanding the implications of their choices on their future careers.
>
> (BDT 2017a)

According to the report, young workers who attended a UTC had a different experience:

> Adam Sullivan, an apprentice electrical engineer at Atkins Global, attended a University Technical

College for his A-levels. He particularly appreciated the project-focussed style of teaching that brought the theoretical work to life.

'We'd learn a concept in maths and then look at how that concept would be used on an actual engineering project. We did the same with chemistry and physics. All the teaching linked back to how a subject would be used within the engineering industry. I think, if teachers can sit down with students and discuss how a subject can be applied, it's a lot better. You understand the application, not just the theory.'
(BDT 2017a)

The report ended with a list of five things which brought the workplace closer to education in UTCs:

1. Employer involvement: employers are actively involved in shaping the curriculum. As a result, UTC students become used to engaging with adults on a professional basis.
2. Embedded careers advice: whether meeting leading engineers, scientists and technologists, getting experience of work or having the opportunity to find out the latest thinking from a leading university expert, students at UTCs are exposed to careers advice every day.
3. Learning in context: at UTCs, teachers help young people to grasp concepts by getting them to apply their learning in technical and practical projects linked to real life.
4. Linking learning to career paths: UTCs link the subjects young people learn to specific STEM career paths with a range of professional qualifications in view.
5. Parity between academic and technical education: UTCs offer a pathway that combines academic, technical and practical learning.
(Summarised from BDT 2017a)

ACCOUNTABILITY MEASURES

The Baker Dearing Educational Trust had always taken the view that UTCs were not set up to deliver the English Baccalaureate, and that Progress 8 was not the best way to measure their success (see chapter 16, above). BDT argued that:

> Progress 8 does not work for UTCs because:
> - Our students only attend UTCs for the last two of the five years covered by the Progress 8 measure. But the progress, or lack of it, made by students at their previous school during Key Stage 3 is wholly attributed to the UTC.
> - Many of the technical subjects on offer at UTCs do not qualify for inclusion in Progress 8.
> - Progress 8 rewards attainment and progress in a small number of core academic subjects linked to the English Baccalaureate. Most UTC students focus on a range of technical subjects and English, maths and science. They tend not to study all of the English Baccalaureate subjects.
> (Baker Dearing Educational Trust, 2018a)

The Department for Education accepted this position in 2018. A note was added to online school performance tables: 'Progress 8 is not the most appropriate performance measure for university technical colleges, studio schools and some further education colleges'. DfE's Permanent Secretary, Jonathan Slater, told the House of Commons Public Accounts Committee that:

> We need a more rounded view of the contribution that the UTC is able to make to the kids who arrive at the age of 14, halfway through their GCSE period. For example, a better measure of the added value of a UTC seems to me to be the extent to which those kids go at the end – after the four to five years they have been at the UTC – into productive employment. It is striking that the average UTC gets 20% of its kids into apprenticeships, which is three times the level of your typical state school. For higher-level apprenticeships, it is four times the level of a typical state school. I think we probably need a more rounded sense of success and failure than we had at the beginning of the programme.
> (HC 691, 18 April 2018, q. 107)

21

ENGINEERING UTC NORTHERN LINCOLNSHIRE

The story of Engineering UTC Northern Lincolnshire in some ways sums up both the problems and the potential of the UTC movement.

The economic case for a UTC in Scunthorpe was based in part on data from the Engineering Construction Industry Training Board which suggested that employers were suffering significant skills shortages, particularly for engineers and technicians. An ageing workforce was expected to boost demand for skilled labour even further in the years ahead.

Humber Renewables and Engineering UTC opened in Scunthorpe town centre in September 2015. Early backers included the University of Hull, North Lincolnshire Council and several leading employers including Able UK, Tata Steel, Total, Centrica and Clugston. Outwood Grange Academies Trust and a nearby further education college, North Lindsey College, also provided support. Other major businesses became more fully involved at a later date, including Phillips 66, Cristal and Singleton Birch.

The application assumed that the UTC would recruit around 170 young people in the first year. Numbers would rise rapidly thereafter, reaching full capacity – 600 places – by the fourth year.

It didn't happen. There was a shortfall in year one and second year recruitment was even more disappointing: by the end of the 2016-17 academic year, there were only 108 students at the UTC. Making matters worse, lower student numbers resulted in a budget deficit, and the UTC was obliged to refund overpayments to the ESFA.

On top of that, students were not getting the experience they had been promised. Employer engagement was extremely limited. Disillusioned students started to think they would be better off completing their studies elsewhere.

Urgent steps were taken to turn things round. An associate assistant principal was appointed to lead on employer projects, and BDT helped facilitate a stakeholder event to boost employer engagement. The trust board was restructured and then in 2017, a new principal was appointed – Marc Doyle – and Graham Thornton (Phillips 66) took over as the chair of governors. A number of arrangements were revised or terminated, including a service contract with Outwood Grange Academies Trust, and the Rodillian Multi-Academy Trust stepped in to provide HR and quality assurance support. Finally, the UTC was renamed Engineering UTC Northern Lincolnshire and given a fresh start.

Even then, the UTC could easily have closed. Student numbers hovered only just above the 100 mark, and finances were tight. The UTC was saved by the transitional funding granted by DfE in 2017, the appointment of Marc Doyle as principal, and the support and challenge provided by a refreshed board of governors. Doyle said:

> I went up through the ranks in a very traditional way. I had five years as a maths teacher, four years after that as assistant head of maths, then head of maths, assistant head for teaching and learning at a small secondary school in Barnsley, did six years as a deputy in Darton, and then moved to Leeds, as associate principal and then principal of a tough secondary. A multi-academy trust took me on to work trust-wide to support new principals. It was good, but by my mid-forties, I thought maybe I should look for something else or I might end up doing this forever.
>
> I came to the UTC and had a look round. I knew it was vulnerable and knew there was a chance it might close. But I could also see so much potential. I saw a great building – it's fantastic and we're very lucky. Some of the facilities are fantastic, too. People say, 'I wish we had things like this when we were young' – a lot of parents say that. But my

first impression was of a sleeping giant that needs injection of enthusiasm and life.

I just thought, right, I like this! It was the first time I'd ever gone for a job and thought, I *really* want this job – I was actually desperate for it. Really, it was the best thing I ever did.

(Doyle, 2018)

Marc Doyle moved from an academy with 1600 students and 170 staff to a UTC with just over 100 students and a staff team of twenty. With growth in student numbers, he increased the staff team to 26, taking the opportunity to restructure the team at the same time. Posts were advertised in the normal way, but Doyle also used links with former colleagues to seek out the specialists he needed. His former trust agreed to second a key member of staff, and agencies helped identify a couple of teachers who were ready to move from their existing schools. A sign of support for the new approach was that no teachers left the UTC at the end of the 2017-18 academic year.

Doyle's very first appointment as principal was a business engagement lead, Carly Boden. He said:

I was very lucky to get Carly. She had worked for Leeds Local Enterprise Partnership but had also been a head of year in a school. And over the last 12 months, with support from our enterprise advisor, Carly has been rebadging our curriculum as a pre-apprenticeship scheme, including things like guaranteed interviews with Phillips 66, British Steel and Cristal. The RAF and Barclays Bank help with mentoring, a solicitor's firm helps with CV writing ... and the University of Hull offers bursaries for students who want to study there when they leave the UTC.

Having Carly makes a huge difference. She pulled together one side of A4 which tells companies the dates when we are doing things and lists things they can help with. She provides information and guidance for employers. She ran a training session for mentors, so they are coming in feeling much more confident and are having meaningful conversations with students.

The result of all that is that one thing often leads to

another. One of our employers has a year 12 student on work placement one day a week. They got hold of our engineering curriculum and they are using the work placement to teach him instead of just sitting him in the workshop one day a week. That's what we want – proper engagement.

(Doyle, 2018)

The board of governors was also refreshed, with a number of senior figures from business, education and the local council showing a keen interest. Marc Doyle said:

The governors believe in this place and the vision is joined up. We're working together on this. It's a better relationship than I've had before: we're a single team working together.

At the same time, I won't let anybody forget that this still functions as a school. The reports that I give them always start with outcomes. It's all very well being told the UTC is fantastic, but the results – they're the things that people see. These kids need to do maths and English, they need to get science GCSEs and high grades in their engineering – we can't just ignore it. And they get that.

(Doyle, 2018)

The largest challenge remained student recruitment. Scunthorpe is a town, not a city. The catchment area is more limited than it would be for a city-based UTC. Furthermore, a demographic dip in 14-16 year-olds meant that only one local school was undersubscribed when Doyle took over as principal of the UTC; and at 16+, the UTC faced competition from two strong and effective colleges, North Lindsey and John Leggott. Indeed, the UTC had only two external applications for post-16 places in 2018-19.

Against that background, and with strong encouragement from North Lincolnshire Council, Engineering UTC Northern Lincs became one of the first UTCs to commit to recruiting at 11. As a stepping stone, the UTC agreed plans to recruit at 13 in 2019-20. Marc Doyle explained:

> We're in a situation in North Lincs at the moment where in years 4 and 5, there's a huge surge in pupil numbers: primary schools are bulging at the seams. The council has done the maths and they know they will soon need another secondary school. The quickest, simplest way is for us to open a key stage 3 building just over the road. We'll have a year of Y7 upstairs [in the existing building] until we've finished the new building, and then they can move out. I think we'd easily attract 120 year 7s, every year. We're already working with two primaries, and we'll be saying to their pupils, if you've got an interest in how things work, if you're interested in science, then come to us.
>
> (Doyle, 2018)

The key stage 3 curriculum would follow national curriculum guidelines, but with a particular emphasis on STEM – or rather, STEAM, where 'A' stands for 'arts': as well as science, technology, engineering and maths, there would be a close connection with an arts centre due to open near the UTC in 2021. The key stage 3 school would naturally feed into the UTC's key stage 4 programmes at 14+, but pupils would have the option to go elsewhere at that age, and the UTC would continue to admit students from other schools who wanted a more technical route at 14+. Early evidence from Leigh UTC, which opened a key stage 3 feeder school (The Inspiration Centre) in 2017, suggested that relations with neighbouring secondary schools started to improve once there was a possibility of two-way transfers of students. Marc Doyle said:

> This opens up a new chapter for UTCs – it's a game changer, really. What I think is really important, is that the UTC movement has to be clear from the outset about what we're trying to do in KS3. Otherwise four or five years down the line, we will be looking back and saying we made some mistakes at the start – the same things that have been said about today's UTCs, in fact.
>
> (Doyle, 2018)

Before the UTC's plans for key stage 3 were fully developed, Ofsted carried out a full inspection, judging overall effectiveness to be good. The report noted that Marc Doyle had 'made a considerable difference in a very short time to the quality of education that the college offers'. Governance was praised, too:

> The new principal is relentless in his efforts to improve the college's effectiveness and quicken its expansion. He acted swiftly and decisively to secure the co-sponsorship of a number of high-profile local and national engineering businesses...
>
> The governors play a vital role in developing and further expanding the UTC. They include members from industry, such as British Steel and Phillips 66, as well as members with specialist educational backgrounds. As a result, they are effective in providing robust challenge to leaders and support the engagement of employers, higher education and [the] local authority.
> (Ofsted 2018a, p.3)

Ofsted also commented favourably on the industry projects devised and co-delivered with the UTC's sponsor employers using high-quality resources, state-of-the-art engineering workshops and the most up-to-date software packages. Inspectors reported that 'pupils are engaged and challenged, and get a good and realistic experience of the world of engineering'. They were also impressed by student attitudes:

> Leaders and staff have high expectations of pupils to behave as young adults. As a result, the vast majority of pupils display mature behaviours and independence ... During the inspection, a number of pupils were keen to engage with inspectors in mature conversations and to shake hands. (Ofsted 2018a, p.5)

Marc Doyle was naturally pleased with the outcome of the Ofsted inspection, including the comment about students shaking hands with the inspectors:

> They shake hands every lesson they go in to; when I walk round the school they will come up to me

and shake me by the hand. The essential thing is the culture we provide. For some kids it's that – the culture – which will make all the difference between success and failure. Here, we emphasize business. The next time you turn a corner in the corridor, the person coming the other way could be your future employer.
(Doyle, 2018)

Student numbers rose from 135 in 2018 to 199 in 2019: still a long way short of capacity, but a sign of progress. Writing on the UTC's website, Marc Doyle said:

Against the odds, the UTC has gone from strength to strength. In danger of closing just over two years ago, with less than 100 students on roll, there was an unerring determination to furnish the local area with skilled professionals. To sit comfortably in 2019 with a Good Ofsted judgement, record Engineering outcomes, 100% of past students in continued employment education and training and a third year where maths and English results are above our stringent national baseline targets, we are delighted to have a school that continues to grow.
(Engineering UTC Northern Lincolnshire, 2019)

22

APPROACHING STEADY STATE

REVIEWS OF GOVERNMENT INVESTMENT IN UTCS

HM TREASURY

In early 2018, HM Treasury officials asked economists at the Department for Education to conduct a value for money assessment of the capital committed to the UTC programme since 2010. BDT staff met the analysts and challenged the basis of their economic model, which used historical data to attach a notional value to different job families with no consideration for trends in future wage returns to qualifications. In particular, the formula assumed that the 'graduate premium' – a measure of increased wage returns achieved by past generations of university graduates – would continue in the future, while those who chose the apprenticeship route would continue to earn less than university graduates. Unless the formula was modified, UTCs looked set to be penalised for helping significant numbers of students to progress into well-paid, fast-track apprenticeships. BDT argued that the net present value of degrees and apprenticeships had already started to change both relative to each other and in absolute terms, and that UTCs would make an outstanding contribution to the UK economy.

BDT met Treasury officials and put it to them that UTCs depended on two propositions:

1. Profound employer engagement with the schools system boosts the quality and number of 18 year olds ready, able and willing to take up jobs and apprenticeships for which there is a skills shortage in the market economy.
2. It is essential to start this technical education at 14 rather than 16.
(Baker Dearing Educational Trust, 2018b)

Perhaps swayed by the first of BDT's two propositions, officials concluded that UTCs provided valuable pathways to apprenticeships, higher education and careers in STEM occupations.

NATIONAL AUDIT OFFICE AND HOUSE OF COMMONS PUBLIC ACCOUNTS COMMITTEE

In 2019, the National Audit Office (NAO) carried out its second investigation into UTCs. The report stressed that 'The investigation does not assess the value for money of the UTC programme' (Great Britain. National Audit Office, 2019, p. 4). Instead, the report provided a commentary on UTCs' progress, financial and educational performance, and plans for improvement.

The NAO noted that between 2010-11 and 2018-19 the government spent a total of £792 million on 58 UTCs, comprising:

- Capital: £680 million
- Pre-opening revenue grants: £62 million
- Transitional revenue funding: £28 million
- Funding to cover UTC budgets: £8.8 million (of which half was expected to be repaid)
- UTC closure costs: £9 million
- Measures to help UTCs improve: £4.5 million
(National Audit Office, 2019, p. 5)

Eight UTCs had already closed: Black Country, Central Bedfordshire, Daventry, Lancashire (Bolton), Greater Manchester, Hackney, Harbourside (Newhaven) and Wigan. South Wiltshire UTC (Salisbury) was due to close in 2020. Two others, Tottenham and Greenwich, had become mainstream academies.

The NAO also noted that few of the 48 remaining UTCs had met their student recruitment targets: on average, UTCs were operating at 45 per cent of their potential capacity. This was a significant factor in the financial challenges faced by UTCs. UTCs as a whole had an accumulated revenue deficit of £7.7 million in 2017/18.

The NAO reported that UTCs performed less well than other secondary schools against the government's preferred measures, such as Progress 8. However, the report also included two important caveats:

> The Department considers that not all its metrics are appropriate for UTCs because of UTCs' technical focus and age range. (National Audit Office, 2019, p. 9)

> The Department's aim is for UTCs to provide clear progression routes into higher education and employment and it therefore considers that student destinations are important performance measures (National Audit Office, 2019, p. 8)

The NAO report was considered by the House of Commons Public Accounts Committee in 2020. The committee also received a briefing document from BDT, which drew attention to additional costs faced by UTCs compared with mainstream secondary schools:

1. More teaching staff to meet a curriculum specification one-third larger at KS4. A UTC KS4 curriculum is 30-33 periods per week vs 25 periods per week at a mainstream school. This is necessary to provide a core curriculum of English, maths, sciences etc. as well as technical specialisms and employer engagement.
2. Higher specialist equipment-related expenditure such as replacement costs, more technicians, and higher energy use.
3. The management of a challenging KS4 cohort. About 40% of the students joining a UTC at age 14 have experienced an atypical education (e.g. previously excluded, home-schooled). They require more support.
   (Baker Dearing Educational Trust, 2020, p.1)

BDT also asked the committee to consider evidence not included in the NAO report:

> The NAO report did not undertake any analysis of the return on investment of a UTC education, however the evidence shows that:
> - The Royal Navy estimates a total salary saving of more than £100,000 over 4 years made possible by accelerating the training of a UTC leaver from 6 years down to 2.
> - BAE Systems believes that UTC students bring a range of skills and experience that will enable the company to offer them a shorter apprenticeship. This could represent a total saving of around £33,000 per apprentice.
>
> (Baker Dearing Educational Trust, 2020, p2)

The Public Accounts Committee took evidence from Jonathan Slater, Permanent Secretary at the Department for Education. Putting the UTC programme into historical context, he said:

> It is striking that the original programme was devised in 2007, when the Department at the time was developing a 14 to 19 curriculum for the whole of the school system. You can imagine that, at the time, this would have been seen as a good way of experimenting with putting that model into practice. The first UTC goes live in 2010; of course, it is then true that, in 2010, the Government's policy changes and moves from 14 to 19 to 11 to 18 academic education. That puts the nature of this programme clearly under increased pressure, but the 2015 Conservative manifesto still said that the objective was to put a UTC within range of every city.
>
> That was before I arrived, but Ministers' ambition at the time must still have been – and why not? – to experiment with seeking to make a technical education curriculum work from 14 to 19 and to see what could be achieved by so doing.
>
> (HC 87, 16 March 2020, q. 10)

Turning to the challenges facing UTCs, the Permanent Secretary said:

> Obviously the key challenge here is the financial challenge, which is driven by the difficulty of getting as many children as had been hoped to switch from one secondary school to another halfway through secondary school, which is obviously a significant challenge. The extent to which they cannot do that as well as they had hoped creates the financial pressure that they find difficulty in coming out of sometimes. That is why we have a series of actions that we take, including putting them into MATs to share the costs, and increasingly turning them into 11 to 19 [schools] …
>
> What triggers [intervention] is a three-year review of the finances of the institution – do they have a deficit today; do they have a plan to get out of it over three years; and can we see whether it is sustainable for them to carry on as they are? If we think it is sustainable, we will put in funding, subject to their agreeing to conditions … if we do not, we close it.
> (HC 87, 16 March 2020, q. 17)

Slater also drew attention to success stories:

> [The NAO] Report and conversations like this [are] absolutely bound to focus on the deficits and the challenges and the finances, but there are some amazing UTCs out there, as Committee members will have seen. We are talking about partnerships with more than 35 universities and 500 employers. Where the model is working really well, you see great examples of employers putting in investment – machinery, equipment, tools, projects – of the sort that the UTC model was designed to secure, and bringing through applicants accordingly.
> (HC 87, 16 March 2020, q. 19)

The Committee's report (HC 87, June 2020) set out four recommendations:

1. The Department should work with those UTCs that have higher occupancy levels to identify and share lessons and good practice for other UTCs that are struggling to attract students.
2. The Department should set clear three-year financial targets for each UTC. At the end of the three-year period, it should be prepared to close UTCs that are not meeting those targets.
3. The Department should, within three months, write to us to explain how it uses data on student destinations to track the performance of UTCs, and what steps it will take to better inform parents about how they can use these data to assess the benefits of a UTC education.
4. The Department should work with UTCs to obtain the information necessary to gain assurance about the value schools are getting from the licence fee they pay to the Baker Dearing Educational Trust, and write to us with its findings within three months.

(HC 87, June 2020, pp. 5-6)

OUTSTANDING ISSUES

FUNDING

As noted previously, the government provided transitional funding worth £200,000 per year, per UTC, until the end of the 2019-20 academic year. This was extended for the academic year 2020-21, but at the lower rate of £100,000 and restricted to UTCs without a key stage 3. Jonathan Slater made it clear to the PAC that this would be the last year of transitional funding.

The government separately announced that 16-19 funding rates would rise in autumn 2020 for all schools and colleges. Other funding decisions were deferred until the next comprehensive spending review, which – at the time – was expected in summer 2020.

Ahead of the comprehensive spending review, BDT set out

several arguments. First, they suggested funding a senior non-teaching post at every UTC to deliver three related functions:

- management of employer engagement
- careers advice and management of student destinations
- management of alumni networks

BDT estimated the cost at approximately £60,000 per annum per UTC and suggested it should be shared equally between employers and the DfE.

Second, BDT put forward ideas for an 'inclusion premium'. Ministers and officials had accepted that an unusually large proportion of key stage 4 UTC students presented challenges, particularly in relation to behaviour and attendance. Many UTCs reported that as many as one in three KS4 students fell into this category, and that they were dedicating considerable staff time to help these students stay on track and minimise the impact of disruptive behaviour on other students. BDT collected data from UTCs which suggested that on average, the additional cost of this support amounted to £130,000 per annum.

Third, BDT asked the DfE to allocate capital funds for the eventual replacement of expensive specialist equipment installed when UTCs were first opened. Charles Parker explained why this was needed: 'We do not have any provision for depreciation or any ability to create sinking funds to renew the equipment. The general annual grant makes no provision for depreciation: that has to change' (Parker, 2018).

Gavin Williamson took up the position of Secretary of State in July 2019. One of his first visits was to UTC Plymouth. He was impressed:

> I worked in manufacturing before becoming an MP, and I know how important it is to have the right skills that this college provides. When you talk to the students you get a true sense of the passion and enthusiasm they have for the skills they are learning here. It is truly inspiring, and we need to be doing more of this.
> (Baker Dearing, 2019c)

BDT hoped the new Secretary of State would respond positively to calls for improved UTC funding. In the event, the Covid-19 pandemic disrupted normal budget processes, including the comprehensive spending review, and at the time this book was completed (July 2020), no decisions had been taken on the long-term funding of UTCs.

STUDENT RECRUITMENT

BDT continued to press for a change in the code of practice on school admissions, arguing that UTCs should be allowed to consider pupils' aptitudes for careers in science, technology, engineering and maths before deciding whether to admit them. DfE resisted the idea, stating instead that local authority letters to parents and the implementation of the Baker clause (see chapter 20, above) would eventually solve the problem of recruiting at 13/14.

In summer 2019, BDT surveyed 25 UTCs to find out whether schools in their areas had taken steps to comply with the Baker clause by allowing providers of technical education to talk directly to all students in years 8, 9 and 11. No UTC reported complete compliance in their area. Most reported limited or no compliance. Some schools ignored repeated requests to speak to their students.

This was backed up by a larger survey conducted by the IPPR later in the year, which found that few schools were complying with the duties set out in the Baker clause (Hochlaf and Dromey, 2019). Only two in five schools (37.6 per cent) had published a provider access statement. Seven in ten (70.1 per cent) of technical education providers said it was difficult to access schools in their areas and fewer than one in three (31.1 per cent) said the situation had improved since the previous year. IPPR concluded that 'Poor compliance is due both to the incentive for schools to retain their pupils, and to the lack of enforcement of the Baker clause' (Hochlaf and Dromey, 2019, pp. 3-4).

Lord Agnew, Parliamentary Under-Secretary of State for the School System, wrote to head teachers of all secondary schools in February 2020, urging them to comply:

Our reforms to technical education are giving young people access to high-quality training and qualifications that match traditional academic routes. You have a crucial role to play in ensuring young people can hear directly from providers of technical education to build up a full picture of the options available to them.

As headteacher, you are under a statutory duty to publish a policy statement setting out details of the opportunities for providers of technical education and apprenticeships to visit your school/s to talk to all year 8-13 pupils, and to make sure the statement is followed. I am grateful that many schools are taking steps to comply with this legislation, commonly known as the 'Baker Clause'. However, too many young people are still not given the chance to learn of different environments open to them and find out if technical education is right for them.

I urge you to take action this term to open your doors to University Technical Colleges, FE colleges, apprenticeship providers and new Institutes of Technology. Now is the crucial moment when so many young people are thinking about their options for September. I recognise it can be challenging, particularly when schools have their own post-16 offer. But we all have a responsibility to support young people to make choices based on their skills, interests and aspirations.

(Agnew, 2020)

In normal circumstances, it might have been possible to assess the impact of Lord Agnew's letter in April or May 2020. In the wake of the Covid-19 pandemic, however, schools had other priorities.

AGE OF TRANSFER

BDT, UTCs and the Department continued to explore the option of recruiting UTC students at age 11, as well as age 13/14 and 16+. Leigh UTC was the first to open a pre-UTC feeder school, recruiting 140 pupils into year 11 in each of its first two years; recruitment directly into year 10 fell to around 65 pupils. Leigh's KS3 curriculum was based on the International Baccalaureate Mid-Years Programme.

During 2018, BDT changed its licensing rules to cover KS3 as well as existing KS4/post-16 UTCs. New KS3 provision would be treated as part of the UTC for funding and legal purposes, but could be presented as 'feeder schools'. It was seen as important for parents to know that their children had the right to move to other local schools at the end of KS3.

All existing UTCs had the opportunity to express interest in opening KS3 feeder schools; twenty did. A particular priority was placed on UTCs in areas where primary schools were experiencing increases in pupil numbers, because this would soon translate into a need for additional places in key stage 3. UTCs were encouraged to base their curriculum plans on Leigh UTC's template, adapting it where appropriate, rather than re-invent the wheel.

By spring 2020, ten UTCs had a year 9 intake (age 13), and two more planned to follow suit in September 2020. At the same time, Bolton, Plymouth and West Midlands UTCs were preparing to open key stage 3 feeder schools with recruitment at age 11.

Simon Connell commented on BDT's change of heart:

> With hindsight, we could have been more pragmatic about the age of entry, because 14 was no longer likely to become a natural age of transfer. We've now got ten UTCs, and two more in September [2020], that start with year 9 as opposed to year 10. It works well with the new Ofsted framework and allows for any progress deficit to be made up over the first year before students start on GCSEs.
>
> And where UTCs want or need to recruit at 11, that's fine too. Leigh UTC now has 400 students in its feeder school. That has clearly made all the difference in terms of recruitment. It's also helped boost the number of girls. Having said that, not all UTCs will move to recruit at age 11, but several intend to do so.
>
> (Connell, 2020)

MULTI-ACADEMY TRUSTS

DfE continued to press almost all UTCs to join Multi-Academy Trusts. Some had already done so, but others had found it difficult to identify suitable partners. BDT provided advice to individual UTCs and kept in close touch with the National Schools Commissioner, Dominic Harrington, and regional school commissioners throughout the process. Moves to establish a UTC MAT in the West Midlands were greeted positively.

By spring 2020, 24 UTCs had joined, or were about to join, a MAT. The DfE hoped a further 15 would join MATs by the end of the year.

TEACHER RECRUITMENT

UTCs did not find it easy to recruiting and retaining all the talented and experienced teachers they needed. Ofsted reports often commented on staff turnover and vacancy rates. Reporting on Bristol Technology and Engineering Academy, inspectors said: 'There has been a high turnover of staff in some subjects ... Some middle leaders are relatively new in post and are not yet ensuring that the quality of teaching in their departments is consistently good' (Ofsted, 2018c, p.3).

UTCs were not alone in finding it difficult to recruit and retain teachers, especially maths and science specialists. A Nuffield Foundation report published in 2018 noted starkly that 'England has had an overall shortage of maths teachers since the 2012/13 academic year and severe shortages since the 2016/17 academic year' (Allen and Sims, 2018, p.10).

However, UTCs faced the additional challenge of recruiting teachers for specialist subjects not widely taught in mainstream schools. Engineering was an area highlighted by BDT's director of education, Ken Cornforth, in a report to the board of trustees (Baker Dearing Educational Trust, 2019a). He estimated that the UTC network needed about 240 specialist engineering teachers in 2019, and that the number would soon rise to over 400; yet most UTCs were already reporting recruitment difficulties. He said:

Recruitment of teachers of engineering is inhibited by:
1. High demand and corresponding salaries for graduate engineers within the industrial sector.
2. A current low overall demand for teachers of engineering in our schools due to its low representation in the secondary school curriculum. The consequential lack of identity has prevented engineering being identified as a subject for teacher training routes. For example, the DfE website "Get into teaching" invites browsers to identify the subject they wish to teach in a secondary school. Engineering is not identified in the menu. Design and technology is identified. However, under that sub-heading, of the 470 entries offering training for design technology, product design, food technology, textiles technology etc, only 10 entries reference "engineering" within their available options i.e. just 2% of those providers.
3. Scholarships and bursaries are available at up to £28,000 for good graduates in maths, computing and physics. Despite these being such key knowledge areas required of the engineer, there is no equivalent incentive available to the teacher of engineering. A lower bursary of £12,000 is however available for teachers of design technology.

(Baker Dearing Educational Trust, 2019a)

The BDT board agreed to lobby for the better identification and recognition of specialist engineering teachers, and for bursaries to be set at the same rate as those offered to new teachers of maths, physics and computing.

That said, UTC students succeeded precisely because they had excellent teachers and teaching assistants – and indeed other staff who went above and beyond expectations to help students.

Ofsted reports made countless references to teachers' professionalism, enthusiasm and commitment, as shown in the following examples:

> Teachers [at UTC Reading] demonstrate impressive levels of subject knowledge, particularly in the college's specialist subjects. This commands the respect of students and ensures high levels of engagement in their learning. (Ofsted, 2015b, p.5)
>
> Teachers' subject knowledge and technical expertise are strong across the academic and vocational curriculum. Pupils [at UTC Sheffield City Centre] have a high level of respect for the expertise and experience staff bring from their work in the engineering and media industries. (Ofsted, 2016b, p.1)
>
> Frequent visits of [Engineering UTC Northern Lincolnshire] staff to industry partners further enhance the knowledge and expertise of teachers. As a result, the quality of teaching and learning across all subjects is strong. (Ofsted, 2018a, pp. 3-4)
>
> The specific needs of disadvantaged pupils [at Energy Coast UTC] are clearly understood by teachers and school leaders. Support for them is excellent and these pupils too make excellent progress. (Ofsted, 2019d, p.7)

T-LEVELS AND PROGRESSION TO HIGHER TECHNICAL QUALIFICATIONS

UTCs played an active part in the development of T-levels, not least by nominating experienced, senior staff to sit on advisory panels working on T-level design and content. Two UTCs, Leigh and London Design and Engineering, were among 52 providers selected to pilot the first T-levels from September 2020. London Design and Engineering UTC subsequently dropped out, but Mulberry UTC was approved to offer the health and science T-level from 2021, and West Midlands Construction UTC (or Thomas Telford UTC, as it became in 2019) was approved to offer the construction T-level from the same year. Another eight UTCs were approved to deliver T-levels from autumn 2022.

Ahead of full implementation, there were still serious concerns about T-levels. For one thing, the DfE anticipated that 180,000 work placements would be needed every year, each lasting between 45 and 60 days: finding this number of placements was expected to be difficult. Secondly – and as noted in chapter 16, above – many UTCs offered a combination of technical qualifications, applied general qualifications and A-levels; for example, a significant number of students took A-level or core maths alongside a technical qualification in science, engineering or technology. This proved especially helpful to students planning to go to university or to start a higher apprenticeship. University admissions staff and employers alike placed great value on a combination of technical qualifications and one or more A-levels. The government appeared intent on removing this option by phasing out alternative technical and applied general qualifications. In their view, students should make a binary choice between A-levels and T-levels, with no half-way house in between. The issue was unresolved when this book was completed.

On the other hand, the Secretary of State for Education, Damian Hinds, said in December 2018 that he was 'determined to properly establish higher technical training in this country' by boosting the availability of level 4 and 5 qualifications:

> I intend to establish a system of employer-led national standards for higher technical education which will be set by employers themselves … we will have the first recognised qualifications in place from 2022 – ready for [the] first T Level students who will just have completed their course. (Hinds, 2018)

TAKING STOCK

NEW UTCS

An application to open a UTC in Doncaster – which had been with the DfE for the best part of two years – was finally approved in June 2018, with recruitment due to start in 2020. The Department for Education said:

> Doncaster University Technical College will train up to 750 13 to 19-year-olds in the latest rail engineering

techniques, as well as coding and 3-D design skills when it opens its doors in September 2020 ...

Parliamentary Under Secretary of State for the School System Lord Agnew said:

"Technology and the world economy are fast-changing, and we need to make sure our young people have the skills they need to get the jobs of tomorrow. There is a clear demand from local businesses for these specialist skills and Doncaster UTC will provide a strong mix of academic and technical-based study that nurtures the talents of all its students."

(Department for Education, 2018)

The DfE announced that a further wave of new free schools would be approved in 2020. There was no separate bidding round for UTCs; instead, UTC applications were considered as part of the wider free school process. Three bids were put forward but none were approved, despite strong support from local employers.

PERFORMANCE

In the 2018-19 academic year, about 13,500 students were enrolled at UTCs. Figures provided to BDT by individual UTCs showed that the number rose to 14,500 in 2020-21.

The Department for Education published school and college performance tables in January 2020 for the 2018-19 academic year (Great Britain. Department for Education, 2020). As before, the principal measures of performance for key stage 4 were Progress 8, entries and achievement in EBacc subjects, and grades achieved in English and maths GCSEs. By 2019, the government accepted that as UTC students generally did not enter the full range of EBacc subjects, Progress 8 and EBacc could not be considered as appropriate performance measures for UTCs.

Turning to post-16 results, the government's preferred measure for most qualifications was progress – that is, how much progress students made between the end of key stage 4 and the end of their post-16 studies, compared to similar students across England. Among 41 UTCs with A-level results reported in the 2019 performance tables, students at four UTCs – Ron Dearing, Oxfordshire, Bolton and Derby – recorded progress well above the national average. Progress was average at 21 UTCs. At UTCs where Tech level results were reported, the average grade was distinction+ at five UTCs; distinction at nine; distinction- at three; merit+ at eight; merit at two; merit- at two; and pass+ at one. In other words, only five out of 30 UTCs achieved below average grades in Tech levels. Finally, the government reported progress made by students retaking English and maths after the age of 16. At UTCs reporting post-16 English and maths results, progress was above the national average in almost all cases (Great Britain. Department for Education, 2020).

A list of UTCs included in the Department for Education's 2019 performance tables (published in January 2020) is provided in an Appendix (below), together with a table of UTC student numbers and characteristics drawn from the same performance tables.

OFSTED

Eighteen UTCs were inspected by Ofsted in the 2018-19 academic year. Ten were graded good and seven as requiring improvement. No UTCs were found to be inadequate. One – Energy Coast – was graded outstanding:

> Leaders [at Energy Coast UTC] promote a consistent culture of high aspiration, high-quality teaching and support and very positive attitudes throughout the UTC. Teachers maintain these standards. Relationships between staff and pupils are excellent. Pupils will confidently discuss their ideas and ask for further explanations when they do not fully understand an unfamiliar concept.
> (Ofsted, 2019d, p.5)

BDT continued to monitor inspections closely. A report to the board of trustees said:

> 2018-19 inspection reports indicate that:
>
> • The quality of leadership and management across UTCs is much improved

- Personal development, behaviour and welfare across UTCs is consistently good
- Inspectors recognised the impact UTCs are making on pupils' engagement, progress and progression
- Inspectors recognised that the UTC curriculum, which develops work skills alongside the acquisition of knowledge, is consistent with high quality destinations.

(Baker Dearing Educational Trust, 2019a)

Ofsted adopted a new inspection framework at the start of the 2019-20 academic year, placing less emphasis on exam results and more on curriculum quality and intent. Ofsted inspectors praised the curriculum at the UTCs inspected in the first half of 2019-20. Elstree UTC moved from 'requires improvement' at its previous inspection to good in all categories. Inspectors were clearly impressed by the curriculum:

> Pupils in this school [Elstree UTC] follow a very special curriculum. They study essential qualifications in English, mathematics and science. Pupils also enjoy great opportunities to learn and experience work in the creative media, arts and entertainments industry.
> (Ofsted, 2019e, p.2)

Liverpool Life Sciences UTC was graded good for a second time:

> Leaders, governors and staff are passionate about helping pupils to become successful and active citizens. The school's motto, 'Every day is an interview', promotes professional standards and high expectations. Many pupils with previous poor experiences of schooling flourish in a caring and supportive environment.
> (Ofsted, 2020a, p.1)

UTC Sheffield City Centre, too, was graded good in all categories:

> Leaders' ambition for pupils who attend the school is high. Leaders have put in place a well-thought-through curriculum. Built around the school's specialist subjects, the curriculum allows pupils to study a broad range of subjects.
> (Ofsted 2020b, p.2)

At Health Futures UTC, inspectors noted a change in focus since the previous inspection, when the UTC had been graded inadequate:

> Since the previous inspection, leaders have made it a priority to make sure that the school has a health and science focus. The curriculum is now more appropriate, with a range of health- and science-related courses.
> (Ofsted 2020c, p.2)

In 2020, Ron Dearing UTC was added to the list of UTCs rated outstanding:

> Ron Dearing UTC has the hallmarks of a school that could be viewed as a guiding light in the educational sector. Pupils attending the school are fortunate to receive an exceptional quality of education … The curriculum is superbly designed to ensure that pupils develop a readiness to learn new content and skills … So far, when leaving the school, all pupils have secured apprenticeships with the school's partner employers, attended university or found alternative employment.
> (Ofsted 2020d, p.2)

THE ROLE OF THE BAKER DEARING EDUCATIONAL TRUST

Earlier chapters described the evolving role of BDT in supporting UTCs. At the outset, the trust devoted most time and effort to helping groups of employers, universities and other interested parties to prepare successful bids to open UTCs. The picture changed when some of the earliest UTCs ran into problems.

To help them, BDT's field team was expanded and strengthened. Working across UTCs in a region, education advisors drew on their own professional experience to identify and share good practice and help principals solve problems. In many cases, the relationship between education advisors and UTCs was strong and effective. Marc Doyle (Engineering UTC Northern Lincolnshire) said:

> I've had a lot of support as well from people like Ken Cornforth [a member of BDT's field team and later Director of Education], who has helped me be bold here. He's been the one to say, come on Marc, you know what you are doing is right – press on.
> (Doyle, 2018)

Jim Wade, principal of the JCB Academy, said:

> The education advisors, the ones I work with, I've found valuable. It's useful to talk to them. It's useful to share ideas and to get their experience coming in, sometimes alongside the DfE education advisor. I think it's evolved really well.
> (Wade, 2018)

However, there were limits to what BDT's advisors could achieve. UTCs were funded by and accountable to DfE. Operating as academies, they had their own boards and appointed their own principals. BDT's only formal leverage was the power to revoke the licence to use the UTC brand and logo – something that could be done only once, as a last resort. In reality, therefore, BDT's influence relied almost entirely on persuasion, cooperation and mutual respect. In a paper for BDT's board of trustees, Peter Wylie reported that:

> In some cases … we have identified problems and been able to broker or support solutions which have produced improvement. In other cases where a diagnosis of issues has been made … the Governors have not been prepared to act. The key problem is that in the absence of a long track record, any review of a school is always going to result in a mix of emerging

evidence and judgements which are open to challenge, not least by those who have entrusted their UTC to a particular leader.
(Baker Dearing Educational Trust, 2018b)

BDT had two trump cards. The first was Lord Baker. His position as a former secretary of state for education and elder statesman gave him unrivalled access to people in high office. As he said himself, people pick up the phone when a former secretary of state makes the call. He had long, deep experience of working in Westminster and Whitehall. In the 1980s, he launched City Technology Colleges, the first schools to be fully independent of local authorities and the precursors of academies. Based on this experience, he knew how to establish independent schools and engage and inspire employers. His arguments for educational opportunity struck a chord with some, while the link between skills and economic prosperity worked with others. In short, it is impossible to imagine that a UTC movement would have got off the ground without Lord Baker's vision and leadership.

BDT's other trump card was the UTC licence. Yes, UTCs were academies; yes, their existence depended on their funding agreements with the secretary of state. But BDT owned the brand. This placed the charity in an exceptionally strong position in talks with ministers, officials and, of course, UTCs themselves. If founders wanted to use the UTC name and logo, they had to sign the licence agreement and abide by a number of immutable principles. This was tested early in the programme, when sponsors of a proposed UTC in Newcastle upon Tyne rejected BDT's requirement that employer and university sponsors should appoint the majority of governors. BDT refused to budge, the licence was withheld and the school opened under another name. It was tested again when a couple of UTC boards decided to change the nature of their provision: by mutual agreement, the licence was withdrawn and UTCs in Greenwich and Tottenham became mainstream schools.

In the five years from 2012/13 to 2017/18, BDT received grants from the Department for Education specifically to work with sponsors planning to open new UTCs. Individual UTCs paid an annual licence fee to BDT, starting at £5000 and rising to £10,000 in 2019/20. The fee contributed towards the support UTCs received from BDT's field team and head office. However, the majority of

BDT's funding came from charitable trusts. In 2015, for example, BDT received £157,000 from DfE, licence fee income amounted to £263,000 and charities donated £901,000. Charities supporting BDT over the years included the Edge Foundation, Gatsby Charitable Foundation, Garfield Weston Foundation, Peter Cundill Foundation, Dulverton Trust and the Michael Bishop Foundation.

At the end of 2018, BDT carried out a strategic review. At that point, only one more UTC was set to open – Doncaster UTC, in September 2020. The question faced by trustees was how best to support a maturing network of UTCs, given that charitable income was expected to fall over the following few years.

A group of core UTC principals met to consider the future. They suggested that BDT's key priorities should be:

1. To achieve a favourable policy and funding environment, in terms of DfE, ESFA and Ofsted
2. Greater public awareness of and support for UTCs, and cross-party political support
3. To preserve UTC distinctiveness and uphold standards, defined by compliance with the UTC licence
4. To facilitate strong and ongoing connectivity and exchange amongst UTCs
5. UTCs should have access to first line advisory support

(Baker Dearing Educational Trust, 2019a)

Writing to all other UTC principals, the core group also made the case for increasing the licence fee to £10,000 from 2019-20. They pointed out that the fee covered a number of benefits and services which would otherwise have to be paid for separately, such as access to GL Assessment's baseline tests for new entrants to UTCs. The licence fee also gave access to Redbourne's data service, which had proved invaluable to UTCs – not least in preparing for Ofsted inspections. Above and beyond these direct benefits, BDT also liaised closely with DfE and other government departments, making the case for the UTC movement and in particular, for fair funding.

BDT's trustees concluded that BDT should become a mutual organization, run by its members. As a first step, three new appointments were made to the board of trustees: Clive Barker, vice chair of governors at the Leigh UTC and Inspiration Academy; David Land, chair of UTC South Durham's board of governors; and Mike Wright, chair of the WMG Academy for Young Engineers trust board. All three also held senior positions in industry. Four other trustees continued in post: Lord Baker, Lord Adonis, Dr Theresa Simpkin and Sir Mike Tomlinson. BDT's central advisory team now comprised Ken Cornforth (director of education) and a specialist in technical education. The number of BDT staff posts was further reduced when Charles Parker retired in 2019: after that, the posts of chief executive and development director were combined, with Simon Connell stepping up to fill both roles. BDT's total income in 2019 was £748,000, derived from UTC licence fees (£358,000), donations (£388,000) and income from investments (£2000).

As UTCs became educationally and financially secure, leaders had more time to support and advise their counterparts in other UTCs. Regular regional meetings, an annual conference and more meetings of principals and specialist teachers cemented a sense of shared identity and purpose: UTCs had become a movement.

Marc Doyle said:

> I've always regarded Baker Dearing as champions of the vision, and I've had a lot of support as well from people like Ken Cornforth, who has helped me be bold here ... This UTC fully buys into what Baker Dearing stands for, and a lot of that stems from my personal beliefs.
>
> I've been involved in a lot of roundtable discussions with other UTCs recently that didn't happen as much before. The fact that we're collaborating more with others – from our point of view, a day a week at Leeds, for a start – means I have a vested interest in achieving success at Leeds UTC as well as here. I have spent time at Durham and Lincoln too, and I meet Sarah [Pascoe] at Ron Dearing UTC once every couple of months. We chat about what's going well, sharing facilities, developing ideas. We're in the middle of a change in the way we do things.
>
> (Doyle, 2018)

It is important to add that there was at all times a strong and dedicated core team of officials within the Department for Education and its agencies whose role included developing policy advice for ministers, overseeing applications to open UTCs, monitoring UTC performance, coordinating links between in-house education advisors and individual UTCs and in the case of the agencies, overseeing UTC finances, including capital investment. There were of course changes in personnel over time, especially when large numbers of civil servants were seconded to work on preparations for Brexit, but the team had a deep knowledge and understanding of the UTC movement's aims, strengths, weaknesses, opportunities and challenges. BDT held them in great respect.

SIMON CONNELL

Transcript of an interview with Simon Connell, Chief Executive of the Baker Dearing Educational Trust.

Back in 2011, I founded an education service called Educate School Services. One thing we did for the free school movement was project management. That led Charles Parker to contact me in 2013 to help with project management at the Medway UTC. I had seen Charles speak at one of the education festivals at Wellington College that summer and was captivated by what UTCs were offering. I jumped at the opportunity to help with Medway, where we did the project management and managed the process for recruiting the principal.

I got to know Charles Parker and Lord Baker better and gradually got more and more involved. I helped the teams behind London Design and Engineering and Ron Dearing UTCs. Then, around the beginning of 2016, I decided to step back from my role at Educate School Services and spoke to Charles about supporting UTCs more broadly. That was when I started to work on a more operational basis for Baker Dearing itself.

From a personal point of view, I guess I've always been a bit of a disruptive thinker: I like to challenge the way things are done and think outside the box. It seemed to me that Kenneth Baker was trying to disrupt the education system with UTCs and that it was the right thing at the right time. Increasingly, young people were leaving school ill-equipped for the real world, whether via university or directly into employment. UTCs were capturing things which matter more than pure academic ability – work-ready skills, project-based learning and all of that.

Growing up in a household where my dad was a part-time cab driver and part-time bricklayer, I've always been impressed by people who've got practical skills. I've got a degree in maths, but I can't change a plug. Society values these things very differently, but having worked with people with practical skills and seen it in my father, I value those skills highly.

I was very fortunate in my own education in that I had a very inspiring maths teacher – without Mr Buckwell I would have been the fourth-generation taxi driver in my family, but instead I got into Cambridge. But it's always struck me that there are lots of people who need a different way of learning. We need to recognize that. We are too 'one size fits all' in this country.

(Connell, 2020)

23

AFTERWORD BY KENNETH BAKER

UTCS: A CENTURY IN THE MAKING?

In the 1890s, new higher grade schools were set up in England to prepare young people for careers in industry and commerce. They combined traditional subjects such as English, maths and history with technical and commercial courses linked to local business needs.

Higher grade schools faced fierce opposition from champions of the grammar school curriculum. Robert Morant, for example, said they offered 'a base, fraudulent, and spurious imitation of education'.

The death knell sounded for higher grade schools when Morant was appointed Secretary of the Board of Education in 1903. A year later, his Secondary Regulations limited external examinations to a handful of subjects: English, maths, science, geography, history, drawing and a foreign language. Nothing else was permitted. Schools which previously taught technical or commercial subjects either adopted the approved curriculum or closed entirely.

The First World War shone new light on technical education. Britain's grammar schools contributed to the officer class, but a lack of technical skills left us exposed in industry and on the battlefield. We had barely started to put this right when the Second World War began.

The 1944 Education Act paved the way for three types of post-war secondary school. Grammar schools continued in the academic tradition, with a strong focus on preparing young people for university and white collar careers. Technical schools taught a combination of academic and technical subjects in preparation for skilled jobs in industry, trade and commerce. Modern schools catered for the majority of children, who were expected to leave school at the earliest opportunity.

Germany adopted this tri-partite model with enthusiasm. Their technical schools were particularly successful and remain popular to this day.

England tells an entirely different story. Only 57 brand new technical schools were opened between 1947 and 1957. Another 300 or so were survivors from the pre-war period, often occupying dilapidated and unsuitable premises and staffed by teachers with limited expertise. Many parts of the country had no technical schools at all.

Technical schools were victims of a peculiarly British snobbery. Parents wanted their children to learn Latin at the grammar school on the hill, not metalwork at the shabby technical school down the road. When comprehensive education gained momentum in the 1960s, technical schools were the first to go.

I very much regretted the loss of technical schools, and one of my first acts as Secretary of State for Education was to launch City Technology Colleges (CTCs). The CTC curriculum was matched to the needs of the modern economy and used computer technology to engage and excite children. CTCs were the first independent state schools, backed by entrepreneurs and captains of industry; they were the forerunners of academies.

I planned a first wave of twenty CTCs. My successors were less keen and the programme was halted in 1994 when there were fifteen of them.

CTCs dropped off the radar for the best part of a decade. Away from the glare of publicity, however, they quietly became some of the best schools in the country.

I also introduced the national curriculum. It was a vital reform which promised a rounded education to all young people, no matter where they lived. If I had my time again, however, I would end the national curriculum at 14.

Shakespeare wrote that 'The fire i' th' flint shows not till it be struck' and the purpose of education is to find that bit of flint – for every child has it – and then to strike a spark from it. Some children are lucky enough to find their passion early, either at school or in their spare time. They know where their ambitions lie and how to achieve them. But they are outnumbered by those who do not.

Offering greater choice at 14 exposes young people to new subjects, ideas and ways of working. It helps them discover hidden talents and shape their future plans.

Tony Blair's government appreciated this. Early pilot programmes gave schools the freedom to offer a wider range of subjects in key stage 4. Ofsted confirmed that pupils and teachers responded enthusiastically and reported improvements in attendance, behaviour, motivation and self-confidence.

The next step was a comprehensive review of the 14-19 curriculum and qualifications, led by the former Chief Inspector of Schools, Mike Tomlinson. His report set out a compelling case for change, giving equal status to academic and vocational qualifications within an overarching 14-19 phase of education.

Tony Blair feared middle class parents would see Tomlinson's proposal as an attack on A-levels, and summarily rejected it. Instead, he announced plans for new vocationally-related qualifications to be known as Diplomas, which would run alongside existing GCSEs and A-levels. Diplomas were rolled out from 2008, but they never caught on as students were taught partly in schools and partly in nearby colleges: this created logistical problems, including synchronizing timetables across institutions. The lesson was that to have a high quality technical education it had to be under one roof, meshing academic and technical learning together.

Meanwhile, the government was shaking up the school system in other ways. Blair's education advisor, Andrew Adonis, took a particular interest in City Technology Colleges and visited almost all of them. At every one, he met enthusiastic leaders, inspiring teachers and motivated students. CTCs' exam results were outstanding and they had superb partnerships with local employers.

CTCs were the spark Adonis was looking for. He proposed replacing failing schools with free-standing academies, each supported by an external sponsor and spurred on by a new mission to succeed.

Blair gave Adonis a seat in the House of Lords, made him Parliamentary Under-Secretary of State for Schools and told him to deliver 200 new academies.

Ron Dearing and I watched all this unfold with great optimism. We were convinced that academies would transform the educational landscape. More than that, we saw an opportunity to set up 14-19 academies specialising in science, engineering and technology. With close links to industry, these academies would deliver Diplomas and apprenticeships through a mixture of classroom study and practical projects.

We spoke to Andrew Adonis, who was very enthusiastic. He said we should start with two and go from there.

Over the following months, Ron and I travelled the country together, inviting employers and universities to share our vision. They leapt at the opportunity to create new routes to rewarding careers in engineering, science and technology. We didn't need to persuade them: we were offering what they wanted.

We called our academies University Technical Colleges: 'university' because each would be sponsored by a university and provide paths to higher education for those students who wanted it; 'technical' to emphasise the specialist curriculum and links with employers; and 'college' to differentiate UTCs from other schools and academies.

With strong support around the country and the 2010 general election on the horizon, we needed firm commitments that the next government would support our plans. We knew we could count on Labour, but we now needed to persuade the Conservative Party to come on board.

Ron Dearing was a respected cross-bench peer with no party affiliations. He used his formidable knowledge and powers of persuasion to great effect both inside and outside the House of Lords. It was a great loss when he died in 2009.

As a former Conservative cabinet minister, I was able before the 2010 general election to present our plans to David Cameron, George Osborne and the future education minister, Michael Gove. George in particular understood that UTCs would be a vital bridge between education and industry, and the creation of twelve UTCs was featured in the Conservative Party manifesto.

Cameron formed a coalition government in 2010 with the support of the Liberal Democrats. George Osborne became Chancellor of the Exchequer and used his first Budget statement to announce funding for at least 24 UTCs. Michael Gove was appointed Secretary of State for Education.

Michael Gove had other educational priorities, from the creation of free schools to changes to the curriculum and examination system, and ending Labour's school building programme. He was less concerned about UTCs and did not set up a unit within his department to promote them. Ron Dearing and I founded the Baker Dearing Educational Trust to promote, develop and support UTCs and to ensure that groups of employers, universities and other supporters prepared applications to open UTCs. We soon exceeded Osborne's promise of 24 and set our sights on at least 50 and in time, perhaps as many as 200.

Building something new is far from easy. The JCB Academy was a success; the Black Country UTC was not. I can sum up the differences quite simply. The JCB Academy opened in a marvellous converted mill with roots in the industrial revolution. Jim Wade was appointed principal fully two years before the academy opened, which gave him ample time to work with employers on a brilliant curriculum, shape his team and recruit the first cohort of students. The Black Country UTC was opened in one corner of a failed secondary school, and it opened in a hurry: there was hardly enough time to recruit a full complement of staff, let alone students. Leadership at Black Country UTC was inconsistent. The first principal left early and there were changes in the board of governors. Student recruitment started to suffer and although sponsors – especially Siemens – put in all the support they could, they were unable to save the UTC from closure.

The JCB Academy had clear, consistent leadership from the very start. The vision shaped by Jim Wade, sponsors, governors and staff has endured. The academy and its students have gone from strength to strength. The academy also became an apprenticeship training provider, supporting several hundred apprentices.

Something the JCB Academy and Black Country UTC had in common was opposition from other secondary schools. Every UTC has experienced it.

Head teachers do not want to lose talented students to neighbouring UTCs, but they are more than happy to wave goodbye to students with a history of bad behaviour, persistent absence or a poor academic record. UTCs found it hard to meet recruitment targets, and in the early years a significant minority of the students who did enrol – especially at the age of 14 – proved challenging. UTC principals did provide extra help with literacy, numeracy and personal development for those students who needed it. Working with employers opened these students' eyes to new opportunities, raised aspirations and boosted motivation. Through mentoring, practical learning and team projects jointly developed and delivered by employers, UTC students gained more than qualifications alone: they developed as people.

A REMARKABLE SUCCESS STORY

Baker Dearing collects information about students who leave their UTC at 16 or 18. When UTC students leave, they are ready for the next step.

At the end of key stage 4, at the age of 16, roughly 40 per cent of students stay on at their UTC to take A-levels and/or technical qualifications. Another 40 per cent move to a different school or college. Significantly, nine per cent of UTC leavers start apprenticeships at 16, compared with a national average of just four per cent.

In 2019, 43 per cent of 18-year-old UTC leavers started courses at a higher education institution and 22 per cent started apprenticeships.

BDT compared these 2019 figures with the government's destination data for young people who completed their studies in 2017 (the latest available figures at that time). There were major differences. Figure 1 shows that 43 per cent of UTC leavers went directly into higher education, compared with a national average (in 2017) of 35 per cent across all state-funded mainstream schools and colleges.

Figure 1: proportion (per cent) of year 13 leavers progressing to higher education – UTCs (2019) and England average (2017)

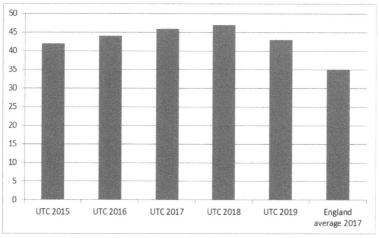

Sources: *UTC data – Baker Dearing Education Trust, based on information supplied by UTCs and UTC students; England average – Department for Education data 2019c.*

More significantly, 77 per cent of UTC leavers going into higher education chose a STEM subject – well above the national average of 46 per cent (see figure 2). Splitting the figures even further, UTC students were considerably more likely to choose an engineering-related course than leavers from other schools and colleges – 41 per cent, compared with just 7 per cent nationally.

Figure 2: proportion (per cent) of year 13 leavers to higher education choosing STEM courses – UTCs (2019) and England average (2018)

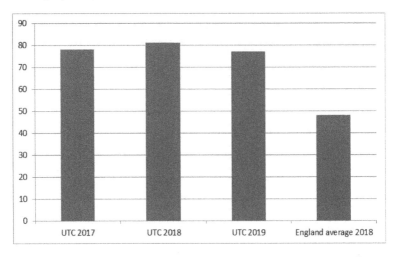

Sources: UTC data – Baker Dearing Education Trust, based on information supplied by UTCs and UTC students; England average – Higher Education Statistics Agency 2020.

Apprenticeship destinations tell a similar story. Figure 3 shows that between one in five and one in four UTC leavers from year 13 go into an apprenticeship, compared with one in ten nationally (10 per cent). Furthermore, a majority of UTC leavers into apprenticeship start higher or degree apprenticeships, compared with under 10 per cent nationally.

Figure 3: proportion (per cent) of year 13 leavers starting apprenticeships – UTCs (2019) and England average (2017)

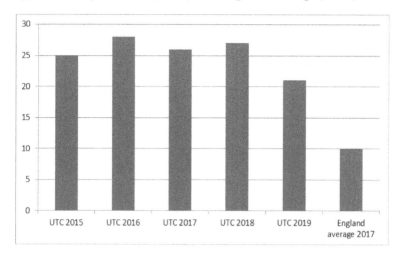

Sources: UTC data – Baker Dearing Education Trust, based on information supplied by UTCs and UTC students; England average – Department for Education data 2019c.

I am immensely proud of the way UTCs help students work towards their goals and in many cases, exceed expectations. UTCs have boosted the prospects of thousands of young people, opening doors to higher education and apprenticeships that some never knew existed – not least in disadvantaged neighbourhoods.

Disadvantage can be measured in various ways. One indicator is the proportion of pupils entitled to free school meals, while neighbourhood deprivation is measured by bringing together several measures in a single index – the 'Income Deprivation Affecting Children Index' (IDACI). When Baker Dearing analysed the destinations of year 13 leavers, we found that many of the highest performing UTCs have high proportions of free school meal students. Similarly, UTCs in high IDACI areas have remarkable track records of helping year 13 leavers progress to higher education. This is social mobility in action.

None of this would be possible without the talent, dedication and sheer hard work of UTC principals, teachers and ancillary staff, supported by governors and sponsors. Five hundred employers are actively engaged with UTCs in 2020, representing a commitment of time and energy on the part of thousands of volunteers over the first ten years of the UTC movement. I pay tribute to each and every one of them.

EDUCATION IS ABOUT MORE THAN MEMORY

Charles Bell coined the phrase 'the intelligent hand' in the 1830s. He understood that the hand guides the brain just as the brain guides the hand. Woodworkers, musicians and chefs rely on touch as much as their sight – in Bell's view even more so, as the eye is easily tricked. The repeated use of touch coaches the brain and stimulates ideas. The American historian, William Rosen, said:

> Modern neuroscience and evolutionary biology have confirmed the existence of what the Scottish physician and theologian Charles Bell called the intelligent hand… We now know that the literally incredible amount of sensitivity and articulation of the human hand, which has increased at roughly the same pace as has the complexity of the human brain, is not merely a product of the pressures of natural selection, but an initiator of it. The hand has led the brain to evolve just as much as the brain has led the hand. (Rosen, 2010, p.36)

The UTC logo is a stylised hand. We chose this design to foster intelligence and imagination by combining knowledge learned in the classroom with practice in the workshop. Our students learn by doing as well as by listening and reading.

Michael Gove did not believe in any of this. He was, in fact, a latter-day Robert Morant. Like his distant predecessor, Gove believed every pupil should take exams in English, maths, science, a foreign language and either history or geography. This combination – the so-called English Baccalaureate – was a re-hash of the 1904 secondary regulations. The only thing missing from Morant's century-old list was drawing.

Gove's ideal classroom consisted of pupils listening attentively as their teacher imparted knowledge. He reformed exams by doing away with almost all project work: grades now depend predominantly on the ability to recall knowledge and write it down in the exam hall. Practical skills count for next to nothing.

The damage done by the English Baccalaureate is clear for all to see. Schools scaled back creative subjects to make room for the EBacc. Figure 4 shows that between 2010 and 2019, GCSE drama entries in England fell by 30 per cent and music entries by 25 per cent. There was an even more dramatic fall in design and technology entries, from 270,401 in 2010 to 89,903 in 2019, a fall of 66 per cent (Joint Council for Qualifications, 2010 and 2019).

Figure 4: number of entries in England in GCSE design and technology, drama and music, 2010 and 2019

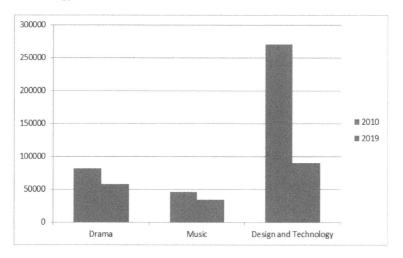

Source: Joint Council for Qualifications, 2010 and 2019.

Music and drama are vital to our lives, our culture and our economy. The same goes for design and technology, which fosters curiosity, creativity, making and doing – working with the hands while engaging the brain.

As a direct result of Gove's reforms, the greatest problem facing our educational system is that it is not providing what industry, business and commerce require.

SKILLS NEEDS, TODAY AND TOMORROW

Professor John Perkins was appointed Chief Scientific Advisor at the Department for Business, Innovation and Skills in 2012. He carried out a review of engineering skills that was published the following year. He wrote:

> The ability to design and make things that the whole world wants to buy remains one of the cornerstones of our economy and one of the most important strategic levers at our disposal to create new growth, new jobs and renewed prosperity. Engineers are the perfect example of the sort of high-skilled, long-term jobs we want to encourage.
> (Department for Business, Innovation and Skills, 2013, p.6)

Presenting clear evidence of skill shortages, Perkins recommended action at every level from GCSEs to professional registration:

> If we are going to secure the flow of talent into engineering, we need to start at the very beginning. We need young people who are technically and academically competent, but who are also inspired by the possibilities of engineering. Starting to inspire people at 16 years old is too late; choices are made, and options are closed off well before then.
> (Department for Business, Innovation and Skills, 2013, p.17)

The Royal Academy of Engineers revisited the Perkins report in 2018 and found little tangible evidence of progress in schools:

> There continue to be shortages of specialist subject teachers in mathematics, physics, computing and design and technology. In many respects the situation has got worse ... Accountability measures on schools in England favour a narrow set of academic subjects

leading to a continued fall in the number of students studying creative, technical subjects such as design and technology that are important in the formation of engineers. (Royal Academy of Engineering, 2019, p. 6)

Engineering continues to remain largely invisible in the school education system, other than in specific provision such as university technical colleges (UTCs), which are currently inadequately supported. (Royal Academy of Engineering, 2019, p. 7)

The umbrella body Engineering UK has repeatedly pointed out that engineering skills are in demand in many sectors of the economy from manufacturing to finance and IT, partly to replace an ageing workforce and partly to support strong growth in key sectors. However, supply has consistently fallen short of demand. Forecasts published in 2019 show that unless we do things differently, there will be a shortfall of between 37,000 and 59,000 unfilled vacancies a year for the foreseeable future (Engineering UK, 2019).

There have been plenty of warnings about engineering skill shortages, but few signs of progress. Figure 5 reveals that –

- Further education colleges in England recorded 139,761 enrolments in engineering and manufacturing technology in 2015/16, but only 123,873 in 2018/19, a fall of eleven per cent.
- In 2015/16, 82,970 students domiciled in England enrolled on university-level engineering and technology courses. The figure rose by five per cent to 87,335 over the following four years: nowhere near enough.
- The number of people participating in engineering and manufacturing technologies apprenticeships fell from 168,542 in 2014/15 to 146,784 in 2018/19 – though there was a welcome shift from low-level intermediate apprenticeships towards advanced and higher apprenticeships over that period.

Figure 5: participation in engineering and technology programmes in further education, higher education and apprenticeships (England), 2015/16 to 2018/19

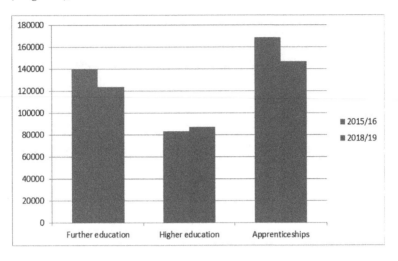

Sources: Department for Education data March 2020a (further education enrolments); Higher Education Statistics Agency 2020 (higher education); Department for Education data 2019b (apprenticeships).

Skills shortages are not confined to engineering. We are in the midst of a digital revolution which is going to change fundamentally patterns of education, training and employment. In 2017 the McKinsey Global Institute forecast that around 20 per cent of current work activities in the UK will be displaced by automation between 2016 and 2030 (McKinsey Global Institute, 2017, p.3). Work will be replaced by intelligent machines not just in manufacturing but in retail, wholesale, logistics, financial services, accountancy, law and many others. Whereas previous waves of mechanisation have eliminated mainly low-skilled jobs, the digital revolution will affect graduate jobs too. This is especially worrying when in 2017, 31 per cent of people with a bachelor's degree already worked in roles that did not require one (Office of National Statistics, 2019).

In 2019 the government's Industrial Strategy Council (ISC) reviewed trends in the labour market and found a superficially encouraging picture: historically high employment rates throughout the country were mirrored by an unemployment rate of under four per cent (Industrial Strategy Council, 2019, p. 9). More worryingly, however, the ISC noted that 'these positive figures mask a significant weakness: many workers do not bring the right bundles of skills to their occupations' (Industrial Strategy Council, 2019, p. 10).

Looking ahead, the ISC appreciated that automation will dramatically alter the skill mix needed both within individual jobs and in the labour market generally. There will be strong demand for the technical skills required to drive the digital revolution forward. Demand will also rise for uniquely human skills including leadership, teamwork, communication, empathy, problem solving and imagination.

Unfortunately, the ISC concluded that skills mismatches will worsen considerably by 2030, with especially acute shortages of specialist skills in science, technology, engineering and maths (Industrial Strategy Council, 2019, p. 28).

Despite all these repeated warnings, the government has been slow to respond. When young people are required to stay in education or training until they are 18, it makes little sense to stick with GCSE originally devised as school-leaving qualifications. Mike Tomlinson was right: we should create an over-arching framework for assessing students as they move through their teenage years, giving as much recognition to technical skills as we give to academic qualifications.

The closest we have come to this so far is the promise of a new suite of qualifications called T-levels, pitched as equivalent to A-levels and appealing to the same age group – 16-19. These stem from a report published in 2016 by an independent panel chaired by Lord Sainsbury (Department for Education, 2016b). Each T-level will be linked to a broad sector of the economy, will be based on the same standards as apprenticeships and will prepare young people for higher apprenticeships or university.

T-levels will be piloted for the first time from September 2020 and will be gradually rolled out over a period of several years. There will eventually be 25 T-levels, including design and development for engineering and manufacturing; engineering, manufacturing, processing and control; maintenance, installation and repair for engineering and manufacturing; and digital production, design and development. UTCs are early adopters of T-levels, but it will be a very long time before they have a significant impact on the country's skill needs.

Separately, an independent panel chaired by Dr Philip Augar was asked to advise the government on post-18 education and training (Department for Education, 2019). The panel strongly recommended strengthening technical education at the level traditionally associated with higher national certificates and diplomas, and more recently foundation degrees:

> England needs a stronger technical and vocational education system at sub-degree levels to meet the structural skills shortages that are in all probability contributing to the UK's weak productivity performance. Improved funding, a better maintenance offer, and a more coherent suite of higher technical and professional qualifications would help level the playing field with degrees and drive up both the supply of and demand for such courses.
> (Department for Education, 2019, p.9)

CORONAVIRUS

Sometimes, events bring a new sense of urgency to our work.

The coronavirus pandemic was a bolt from the blue. In an effort to limit the spread of Covid-19, the government imposed a lockdown affecting almost all parts of public and private life. We depended on the skills, compassion and professionalism of health and care staff to see us through the peak: not just doctors, nurses and paramedics, but countless technicians and experts working long hours in hospitals and laboratories throughout the land. Engineers, teachers and students repurposed 3D printers and laser cutters to make much-needed personal protective equipment. Scientists worked at record speed to identify possible treatments and develop vaccines.

Suddenly, too, we discovered new ways of working remotely using Zoom, Teams, Skype and other cutting-edge applications. We did our shopping on the internet and streamed films, TV and games to keep ourselves entertained. Schools, colleges and UTCs made sure children and young people could learn from home. And throughout the crisis computer scientists, programmers and technicians made sure none of the technology failed in our hour of need.

It is hard to think of another moment in peacetime when science, engineering and technology have been more important to our lives.

The pandemic brought with it a sharp, deep and potentially long-lasting recession. Despite loans and subsidies, jobs were lost across the UK economy. Sadly, many apprenticeship places were cancelled.

The government set about rebooting the economy at the first opportunity, announcing massive investment in capital projects, skills and green technologies to combat climate change.

All of the inquiries, reviews and reports of the last decade said we needed to turbo-charge education and training for science, technology and engineering. Could it be that Covid-19 will turn out to be the catalyst we need to make it happen? Will we look back on 2020 as the start of a new golden age for STEM education?

UTCS: THE NEXT TEN YEARS

The years after Covid-19 are going to be very challenging for there will be high levels of youth unemployment. As the country begins to recover, we should not return to the low-skilled, low-paid economy of the last twenty years. It should be essential to improve substantially technical, practical education leading to employability skills. Our young people must not acquire just academic knowledge, they should be given the skills of problem-solving, working with their hands, designing things on computers, using 3D printers – and they should be encouraged to be entrepreneurial. Technical education should be reintroduced into schools below the age of 16. I would like to see the unique offering that UTCs provide extending into 200 UTCs by 2030.

Tony Blair said, 'Education, education, education'. We should be saying, 'Skills, skills, skills'.

It has taken ten years for UTCs to become part of our education system and that has been achieved with little ministerial enthusiasm and a Department for Education that has at times been reluctant and diffident. If I had not held the post of Secretary of State for Education and not known how the Department worked, I think the initiative that Ron Dearing and I launched would have seen UTCs transformed into mainstream schools.

Any new experiment must have a bedding-down period. We have shown how good UTCs are for they are now receiving Ofsted grades of 'outstanding' and 'good'. We are proud of the destinations data of our students: 43 per cent go to universities where most study STEM courses; 25 per cent become apprentices; and we are transforming the life chances of young people who were disengaged at their previous schools and wanted a fresh start. We have the support of over 500 companies who are closely involved in the UTC curriculum, which they see as a direct benefit to their own futures.

UTCs have had detractors and outraged opponents, but we have survived because our students know there is a need for high quality technical education and that it will lead to a better life for them.

So, let me end by recording what some of our 2020 cohort said in a survey of over 300 UTC students – they, after all, are the best judges of our success:

To be honest my UTC is the best place to study if you are looking for a future career and good education.

Go for it! Best decision you will ever make to develop skills and confidence, as well as knowledge about the job sector and industries in the area.

If you are passionate about a career in engineering, it can be both career-defining and life-changing … if you work hard you can get some truly phenomenal skills and moments from your time at a UTC.

I would definitely recommend going to a UTC as it allows you to gain good contacts with national and local companies. I also had a fantastic work experience placement with a company through a contact tailored to my interests.

It is the best thing I have ever done. I had a fantastic time with the UTC. It is brilliant!

24

JODIE'S STORY

Sheffield student Jodie Howlett has begun a nine-week program at the International Space University after a degree in engineering. Laura Drysdale reports.
Since its founding in 1987, the International Space University (ISU) has been at the core of the global space community.

Yesterday, Sheffield Hallam University student Jodie Howlett was one of just six people across the UK who joined the centre's renowned Space Studies Program.

(Drysdale, 2019)

Jodie Howlett spoke to the author while taking part in the European Space Agency's space study programme in Strasbourg and shortly after graduating from Sheffield Hallam University with first class honours in mechanical engineering.

I didn't feel I was working to my full potential at my previous school in Burton on Trent. I was in the middle set for everything. I knew I really liked science and maths and being creative, but I didn't have a clear vision of what I wanted to do for a career.

My mum heard about the JCB Academy on the radio and suggested I take a look at it. We had a tour on one of the open days and a couple of induction sessions at the JCB business headquarters before starting at the Academy. I liked it that they offered engineering and business and thought that it would be a really good place to study. It appealed to me that we would get to work with industry. Having said that, it hadn't actually opened when I applied, so I took a leap of faith really!

My impression on the first day was that everything was extremely professional and high-profile. The media were there covering the opening of the Academy and interviewing students. The professionalism carried through into the curriculum, thanks to the staff who worked hard to make a business-orientated environment and to treat us as adults, which I really enjoyed.

The journey from home to the Academy was about an hour and obviously, the same again in the evening. It took a while to get used to the long school day: everyone was very tired for the first few weeks! But it was worth it. I loved going to the Academy.

The contact we had with employers was the most enjoyable thing for me. In fact, I'd say it was the most valuable aspect of the curriculum. We had numerous trips out to visit sites and people came into the Academy from a variety of companies. They came from a wide mixture of engineering backgrounds, which helped provide a good overview of the different areas we could work in. You just wouldn't get that at a normal school.

I particularly remember the Rolls-Royce challenge where we had to design and manufacture a small piston – a fuel pump for a jet engine. It involved a mix of hands-on practical work and planning, quality, preparing a business case, minimising waste and things like that.

We got better at teamwork over time as we got to know each other and understand each other's strengths. For myself, I sometimes found it harder to get involved in the more practical side of the projects, but we worked together well – we were all willing to support one another.

During key stage 4, I learned that I was very analytical person, with good attention to detail. The work I was producing was to a high standard. I was very good at spotting small mistakes and always looked for ways to improve the way we did things. That's definitely part of the skills set needed in engineering, which made me more comfortable choosing engineering in sixth form rather than business.

I found sixth form a step up in terms of what was expected of us. We were treated a lot more like adults and given greater responsibility. It felt less and less like a school environment. For my extended project qualification, I was looking at how GPS could be used to control the speed of a car. I was interested in vehicle safety and preventing people going faster than the speed limit. I made a small prototype using a microcontroller as well as researching and writing about the idea.

We all had a mentor in key stage 4. We discussed things like attendance, grades, whether there was anything we were struggling with, how we were finding different classes – just an overview of life at the school. We didn't have a specific mentor in sixth form, but I built up a good relationship with my engineering teacher, Paula Gwinnett, and in a way she was a mentor to me. She taught me engineering all the way through my time at JCB Academy and I'm still in touch with her today. She has been really supportive, written references for me, put me in touch with people in industry and was a real inspiration – the first role model I had in engineering. She brought lots of experience from her previous career in engineering to the classroom. She was very good at showing the context for what you are learning and how it would apply in real life.

It's really important that girls should be able to see themselves in someone else's shoes. But there's a gap in the middle – 'I'm here, my role model is there. I don't know how to get from here to there'. We need to help girls understand exactly what they have to do to get from here to there. They need to see it as something that is achievable.

In my experience, boys tend to like making things and being very hands-on, whereas the women I've worked with are generally motivated by the impact they can have through their work – the social and environmental impact in particular. In other words, it's not just the hard, here-and-now engineering, but the difference you're making to the world. And another thing: engineering isn't about working in isolation. There's a lot of teamwork involved. Paula used to tell stories of when she worked at Procter and Gamble, leading teams and working all over the world. I found that really inspiring.

As I started to look at options for the future, I was already keen on aerospace engineering, but at the time I was choosing university courses I wasn't entirely sure if that's what I wanted to do. I didn't really know any engineers before I went to JCB Academy – there was no family history of engineering – so Paula talked to me about different engineering courses. She studied mechanical engineering at university and introduced me to the idea of becoming a chartered engineer.

I did three different summer schools at universities whilst I was still in the sixth form. In the space of a single summer, I went to Cambridge, Warwick and Birmingham and spent a week at each. I funded that by working Saturdays. I was keen to put myself in a good position when applying for university. I looked at general engineering and physical sciences, which helped me make my mind up. I opted for mechanical engineering to keep my options open, knowing that I could move into aerospace later if I wanted to.

Academic studies haven't always been easy for me. I did really well in GCSEs but I struggled with A-levels. I didn't get the grades I needed for my first choice of university – in fact, I had to go through clearing to get a place at Sheffield Hallam University.

At university, I noticed some differences when I compared other students' experience with my own. JCB Academy had geared me up to be very proactive. As soon as I started, I was already looking ahead to the year in industry because I saw that as one way to become as employable as possible. I found out I could do summer internships and in my first year I applied to over 20 companies. I was accepted by Siemens in Oxford. For me, that was one of the things that most helped me to succeed and to stand out at university. Other students didn't approach it in the same way: many of them didn't do a year in industry – they weren't really interested in the idea. I have classmates who have still not applied for graduate-level jobs.

I also felt better prepared for working in teams. I was aware that that was something companies look for, both for internships and permanent positions. Teamwork skills really helped in job applications and assessment centres: I knew how to interact with others and give them encouragement and support. I was ready to take the lead when necessary. I knew how to input into conversations and discussions. All that was seen as a strength. Working in teams at the JCB Academy really helped with that.

Last year, I went to a study conference about aerospace engineering. The funny thing is, I wasn't meant to go – I was planning a kayaking trip – but we got lots of heavy snow so I went to the conference instead. I bumped into one of my role models there, who works in the rail industry. I said, 'I didn't know you were interested in aerospace,' and she said 'yes, I went on a space study programme after finishing university'. She introduced me to the programme and encouraged me to apply. And now, here I am in Strasbourg!

The programme is mostly funded by the European Space Agency, but I also got a grant from the Royal Academy of Engineering. The whole programme is incredible and I feel very lucky to be here. It's very intense: you meet people from different cultural backgrounds, we work until nine, sometimes ten in the evening, some days we have lectures after dinner. We get to study every different aspect of space, not just engineering. We cover law, business management – even space medicine, which is the aspect I've chosen to specialise in while I'm here.

I also applied for a one-year graduate traineeship with the European Space Agency. I completed some online tests, did a recorded video interview and then had a face-to-face Skype interview before being offered the place. I will be based in Darmstadt in Germany, starting in September. My role will be in process performance analysis and quality management. The programme helps prepare people for career opportunities in industry – with Airbus, for example, or wherever there are suitable aerospace opportunities. Very often, after five or 10 years, people then return to the space agency. In ten years' time, I would like to be working in mission control at the ESA, ideally as a flight director.

There are four main differences between the JCB Academy and an ordinary school. Project-based learning, industry engagement, the professional environment and the amazing staff. Why wouldn't you go to a UTC? It has all the ingredients of an ordinary school – you still do the regular qualifications – but there's so much extra to be gained through the experience. As long as you're willing to work hard, there's nothing to lose.

If I was talking to someone thinking about moving to a UTC, I would say this. Going to a UTC doesn't, in itself, guarantee a place with a company or on a particular career path: you get out what you put in. Resilience and a positive attitude make a big difference, too. But if you take that leap, take that chance, the results can be incredible – life changing. Going to JCB Academy has set me up for life.

(Howlett, 2019)

APPENDIX

When the Department for Education published performance tables for the academic year 2018-19, they included data from 49 UTCs (Department for Education, 2020). Table 4 lists UTCs alphabetically, showing their locations and specialisms.

In total, over 13,500 students were enrolled at UTCs in 2018-19. A breakdown of student numbers and characteristics is provided in table 5. (UTCs which closed in 2019 and 2020 have been excluded.)

Table 4: UTCs included in Department for Education performance tables published January 2020, with location and specialisms

University Technical College	Location	Specialisms
AldridgeUTC@MediaCity	Salford	Creative and digital industries, design
Aston University Engineering Academy	Aston	Engineering, science
Bmat Stem Academy	Harlow	Computing, science, engineering
Bolton UTC	Bolton	Health sciences, engineering
Bristol Technology and Engineering Academy	Bristol	Engineering, computer science
Buckinghamshire UTC	Aylesbury	IT and computing, construction
Cambridge Academy for Science and Technology	Cambridge	Biomedical science, environmental science, technology
Crewe Engineering and Design UTC	Crewe	Engineering, manufacturing, design
Derby Manufacturing UTC	Derby	Manufacturing, engineering
Elutec	London	Product design, engineering
Energy Coast UTC	Workington	Construction, engineering, energy

Engineering UTC Northern Lincolnshire	Scunthorpe	Engineering, renewables
Greater Peterborough UTC	Peterborough	Engineering, built environment
Health Futures UTC	West Bromwich	Health care, health sciences
Lincoln UTC	Lincoln	Engineering, science
Liverpool Life Sciences UTC	Liverpool	Science, healthcare
London Design and Engineering UTC	London	Design, engineering
Mulberry UTC	London	Digital technology, healthcare and medical services
Ron Dearing UTC	Hull	Digital technologies, mechatronics
Scarborough UTC	Scarborough	Engineering, computer science, cyber security
SGS Berkeley Green UTC	Berkeley	Digital technology, advanced manufacturing, cyber security
Silverstone UTC	Towcester	High performance engineering, business and technical events management
Sir Simon Milton Westminster UTC	London	Transport engineering, construction
South Bank Engineering UTC	London	Engineering, building systems engineering, medical engineering
South Devon UTC	Newton Abbot	Science, engineering, environmental science, computer science
South Wiltshire UTC	Salisbury	Science, engineering
The Elstree UTC	Borehamwood	Multimedia, production arts, digital technology and communication for entertainment industries
The Global Academy	London	Creative, technical and broadcast, digital media
The JCB Academy	Rocester	Engineering
The Leigh UTC	Dartford	Computer science, engineering
The Watford UTC	Watford	Computer science, travel and tourism, hospitality, event management
Thomas Telford UTC	Wolverhampton	Construction, ICT in the built environment
UTC Heathrow	Northwood	Aviation engineering, engineering
UTC Leeds	Leeds	Engineering, manufacturing
UTC Norfolk	Norwich	Advanced engineering, energy
UTC Oxfordshire	Didcot	Life sciences and physical sciences, engineering
UTC Plymouth	Plymouth	Science, engineering

UTC Portsmouth	Portsmouth	Electrical and mechanical engineering, advanced manufacturing
UTC Reading	Reading	Computer science, engineering
UTC Sheffield City Centre	Sheffield	Engineering, creative and digital media
UTC Sheffield Olympic Legacy Park	Sheffield	Health and sports sciences, computing
UTC South Durham	Newton Aycliffe	Science, engineering, environmental science, computer science
UTC Swindon	Swindon	Engineering
UTC Warrington	Warrington	Engineering, science
UTC@harbourside	Newhaven	Marine engineering, environmental engineering
Waterfront UTC	Chatham	Engineering, construction, design
Wigan UTC	Wigan	Engineering
WMG Academy for Young Engineers (Coventry)	Coventry	Engineering, digital media
WMG Academy for Young Engineers (Solihull)	Solihull	Engineering, science

Table 5: UTC student numbers and characteristics, 2019 (ordered by total student numbers, largest to smallest)

Name	Head-count 2019	% Pupils with SEN support	% Girls	% Boys	% free school meals last 6 years
The JCB Academy, Rocester	697	17.9	22.2	77.8	11.0
Aston University Engineering Academy	608	6.9	24.0	76.0	50.7
The Leigh UTC, Dartford	481	13.7	24.5	75.5	34.2
Liverpool Life Sciences UTC	476	7.1	57.8	42.2	57.3
UTC Sheffield City Centre	460	13.0	25.2	74.8	27.9
Silverstone UTC	456	5.9	16.4	83.6	16.6
UTC Reading	453	14.6	15.2	84.8	15.5
AldridgeUTC@MediaCityUK, Salford	426	8.9	41.3	58.7	40.1
Ron Dearing UTC, Hull	422	3.1	14.0	86.0	31.0
London Design and Engineering UTC	404	6.9	25.0	75.0	47.7
UTC Sheffield Olympic Legacy Park	403	7.7	48.9	51.1	28.0
The Elstree UTC	392	29.8	63.0	37.0	35.0
UTC Oxfordshire, Didcot	362	25.4	21.3	78.7	18.5
WMG Academy for Young Engineers, Coventry	352	9.4	21.3	78.7	23.1
UTC South Durham, Newton Aycliffe	341	16.4	17.3	82.7	39.6
University Technical College Leeds	335	6.0	17.3	82.7	31.1
WMG Academy for Young Engineers, Solihull	324	10.5	15.4	84.6	36.2
UTC Portsmouth	319	19.4	20.7	79.3	21.2
The Global Academy, Hayes	297	26.6	55.2	44.8	28.0
Energy Coast UTC, Workington	285	0	34.7	65.3	35.1
University Technical College Norfolk, Norwich	266	13.9	18.4	81.6	23.3
Health Futures UTC, West Bromwich	265	10.2	74.7	25.3	53.6
Greater Peterborough UTC	261	24.9	17.6	82.4	23.5
UTC Warrington	258	5.8	22.1	77.9	28.5

Sir Simon Milton Westminster University Technical College	246	19.5	17.5	82.5	55.5
Bristol Technology and Engineering Academy	240	16.3	21.7	78.3	24.4
Derby Manufacturing UTC	235	10.6	16.6	83.4	33.3
SGS Berkeley Green UTC	232	28.0	11.6	88.4	23.7
Mulberry UTC, London	229	13.1	80.8	19.2	58.5
South Bank Engineering UTC, London	220	13.2	16.4	83.6	45.3
Crewe Engineering and Design UTC	218	15.6	22.5	77.5	30.9
Cambridge Academy for Science and Technology	217	27.6	35.5	64.5	15.1
Scarborough University Technical College	199	7.0	19.1	80.9	20.8
Thomas Telford UTC, Wolverhampton	189	24.3	21.7	78.3	40.0
Lincoln UTC	186	9.7	23.1	76.9	26.3
Buckinghamshire UTC, Aylesbury	185	24.9	10.8	89.2	32.8
UTC Heathrow	182	8.2	9.9	90.1	35.1
The Watford UTC	174	7.5	40.8	59.2	36.7
Waterfront UTC, Chatham	170	22.9	20.6	79.4	29.2
South Devon UTC, Newton Abbott	165	26.1	18.2	81.8	36.4
Bolton UTC	158	5.7	60.8	39.2	43.0
North East Futures UTC, Newcastle	143	8.4	39.9	60.1	41.7
Elutec, Dagenham	139	8.6	18.0	82.0	33.0
Engineering UTC Northern Lincolnshire, Scunthorpe	130	27.7	10.8	89.2	37.5
UTC Swindon	121	21.5	12.4	87.6	31.1
UTC Plymouth	80	22.5	11.3	88.8	34.4
Bmat Stem Academy, Harlow	50	6.0	36.0	64.0	0.3
England state-funded secondary schools	3,327,970	10.8	49.8	50.2	27.7

(Source: Department for Education, 2020)

REFERENCES

Adonis, A. (2011). 'A century late, the technical school is with us at last', *The Times*. 8 January. Available at: https://www.thetimes.co.uk/article/a-century-late-the-technical-school-is-with-us-at-last-5x6zcprmq8h. (Accessed: 22 May 2020).

Adonis, A. (2012). *Education, Education, Education* [ebook; Nvidia Shield Tablet]. London: Biteback Publishing. Viewed 22 January 2018. Available at: Amazon.co.uk

Agnew, T. (2020). *Schools' compliance with the 'Baker clause'*. Letter from Lord Agnew, Parliamentary Under-Secretary of State for the School System, to school head teachers in England, 7 February 2020.

Allen, R. and Sims, S. (2018). *How Do Shortages of Maths Teachers Affect the Within-School Allocation of Maths Teachers to Pupils?* London: Nuffield Foundation. Available at: https://www.nuffieldfoundation.org/wp-content/uploads/2018/06/Within-school-allocations-of-maths-teachers-to-pupils_v_FINAL.pdf (Accessed: 3 June 2020).

Anon. (2011). 'Crying Wolf', *The Times*, 8 March 2011. Available at: https://www.thetimes.co.uk/article/crying-wolf-5k7cfp2l8tm (Accessed: 22 May 2020).

Archer, L., DeWitt, J., Osborne, J., Dillon, J., Willis, B. and Wong, B. (2013). '"Not girly, not sexy, not glamorous": primary school girls' and parents' constructions of science aspirations'. *Pedagogy, Culture & Society* [online]. Vol. 21, No. 1, 171–194. Available at: http://dx.doi.org/10.1080/14681366.2012.748676 (Accessed: 10 January 2019).

Bailey, J. (2018) interviewed by the author, 23 January 2018.

Baker Dearing Educational Trust (2009a). *Charitable objects*. Available at: http://apps.charitycommission.gov.uk/Showcharity/RegisterOfCharities/RemovedCharityMain.

aspx?RegisteredCharityNumber=1129510&SubsidiaryNumber=0 (Accessed: 7 February 2018).

Baker Dearing Educational Trust (2012). *Proposal for a Professional Tec Bac.* [Unpublished item].

Baker Dearing Educational Trust (2013a). Press release, 8 July [unpublished item].

Baker Dearing Educational Trust (2013b). *Summaries of Interviews with Candidates for Duke of York Awards.* [Unpublished item].

Baker Dearing Educational Trust (2014a). *Survey of female students.* [Unpublished item].

Baker Dearing Educational Trust (2014b). *UTC capacity and cohort numbers 2014-15 academic year.* [unpublished item].

Baker Dearing Educational Trust (2015a). *Student Recruitment Guide: A Guide to UTC Student Recruitment.* London: Baker Dearing Educational Trust.

Baker Dearing Educational Trust (2015b). *Analysis of UTC exam results and student destinations 2015.* [Unpublished item].

Baker Dearing Educational Trust (2016). *Report to Baker Dearing Educational Trust Board of Trustees,* July. [Unpublished item].

Baker Dearing Educational Trust (2017a). *From School Work To Real Work: How Education Fails Students In The Real World.* London: Baker Dearing Educational Trust.

Baker Dearing Educational Trust (2017b). *Report to Baker Dearing Educational Trust Board of Trustees*, February. [Unpublished item].

Baker Dearing Educational Trust (2017c). *UTC students experience life as a Royal Navy Engineer* [online]. 24 October. Available at: https://www.utcolleges.org/utc-students-experience-life-royal-navy-engineer/ (Accessed: 29 May 2020).

Baker Dearing Educational Trust (2017d). *An Introduction to UTC Governance*, May. [Unpublished item].

Baker Dearing Educational Trust (2017e). *Report to Baker Dearing Educational Trust Board of Trustees*, July. [Unpublished item].

Baker Dearing Educational Trust (2018a). *Performance Measures Statement by Charles Parker*, January. Available at: https://southdevonutc.org/2017-performance-measures/ (Accessed: 29 May 2020).

Baker Dearing Educational Trust (2018b). *Report to Baker Dearing Educational Trust Board of Trustees*, March. [Unpublished item].

Baker Dearing Educational Trust (2019a). *Report to Baker Dearing Educational Trust board of trustees,* June. [Unpublished item].

Baker Dearing Educational Trust (2019b). *Insights – UTC Leaver*

Destinations 2019. Available at: https://www.utcolleges.org/wp-content/uploads/2020/01/BD-insights-V11.pdf (Accessed: 2 June 2020).

Baker Dearing Educational Trust (2019c). *The Secretary of State for Education Makes a Special Visit to UTC Plymouth* [online]. Available at https://www.utcolleges.org/the-secretary-of-state-for-education-makes-special-visit-to-utc-plymouth/ (Accessed: 1 July 2020).

Baker Dearing Educational Trust (2020). *Written Evidence to the Public Accounts Committee*. Available at: https://committees.parliament.uk/writtenevidence/732/html/ (Accessed 2 June 2020).

Baker, K. (2009). *Address at the Service of Thanksgiving for the Life and Work of Ron Dearing*, 7 July. [Unpublished item].

Baker, K. (2011a). Personal diary entry. 24 January. [Unpublished item].

Baker, K. (2011b). Personal diary entry. 10 March. [Unpublished item].

Baker, K. (2011c) interviewed by the author, 3 August 2011.

Baker, K. (2012). Personal diary entry. 21 June. [Unpublished item].

Baker, K. (2013). 'The Duke of York to Launch New Award for Technical Education' [press release]. Baker Dearing Educational Trust, 8 July.

Baker, K. (2014a). Personal diary entry. 10 February. [Unpublished item].

Baker, K. (2014b). Personal diary entry. 24 February. [Unpublished item].

Baker, K., (2016a). 'We must not freeze parents out of the academies revolution'. *Daily Telegraph*, 13 April. Available at: https://www.telegraph.co.uk/news/2016/04/13/we-must-not-freeze-parents-out-of-the-academies-revolution/ (Accessed: 13 February 2019).

Baker, K. (2016b). *The Digital Revolution: the Impact of the Fourth Industrial Revolution on Employment and Education*. London: Edge Foundation. Available at: https://www.edge.co.uk/sites/default/files/documents/digital_revolution_web_version.pdf (Accessed: 13 February 2019).

Baker, K. (2016c). Personal diary entry. 14 August. [Unpublished item].

Baker, K. (2016d). Personal diary entry. 20 September. [Unpublished item].

Baker, K. (2017a). Personal diary entry, 1 February 2017. [Unpublished item].

Baker, K. (2017b). Letter to the editor [untitled]. *The Times*, 15 February.

Baker, K. (2019). 'The official UTC figures don't give the full picture'. FE Week, 3 February. Available at: https://feweek.co.uk/2019/02/03/the-official-utc-figures-dont-tell-the-full-story/ (Accessed: 20 May 2019).

Baker, M. (2005) 'Why Tomlinson was turned down', *BBC News*, 26

February. Available at: http://news.bbc.co.uk/1/hi/education/4299151.stm. (Accessed: 12 February 2018).

Bhattacharyya, K. (2017). 'Super-selective grammars and technical schools'. *The Times*, 13 February.

Bell, D. (2018) interviewed by the author, 23 January 2018.

Best, F. (2014). *University Technical Colleges: Opening Up New Opportunities For Girls*. Bradford: WISE Campaign. Available at: https://www.raeng.org.uk/publications/reports/university-technical-colleges-opening-up-new-oppor (Accessed: 26 March 2019).

Best, F. (2018) interviewed by the author, 18 April 2018.

Black Country UTC (2010). *The Black Country UTC: Briefing Note*. December 2010. [Unpublished item].

Boles, N. (2015). Letter to Principals and Chairs of Governors of University Technical Colleges, 15 October. [Unpublished item].

Boles, N. (2016a). Letter from Nick Boles to Lord Baker, 2 March. [Unpublished item].

Boles, N. (2016b). Letter to Principals and Chairs of Governors of University Technical Colleges, 3 March. [Unpublished item].

Bragg, V. (2011). Email to Baker Dearing Educational Trust. 24 April. [Unpublished item].

Brooks, G. (2018) interviewed by the author, 10 May 2018.

Byrne, J. (2015). Transcript of a presentation at an Edge Foundation seminar held in November 2015. [Unpublished item].

Clarke, L. (2019) interviewed by the author, 4 July 2019.

Connell, S. (2016). *A Guide to Strategic Financial Planning for UTCs*. London: Baker Dearing Educational Trust. [Unpublished item].

Connell, S. (2020) interviewed by the author, 2 April 2020.

Conservative Party (2009). 'Gove: A New Generation of Technical Schools' [press release]. 5 October.

Conservative Party (2015). *Strong Leadership, A Clear Economic Plan, A Brighter, More Secure Future* [the Conservative Party manifesto]. London: the Conservative Party. Available at: http://ucrel.lancs.ac.uk/wmatrix/ukmanifestos2015/localpdf/Conservatives.pdf (Accessed: 27 May 2020).

Cornforth, K. (2018). 'UTCs pioneer next generation technical education'. *The Engineer*, 31 October. Available at: https://www.theengineer.co.uk/utcs-new-generation-technical-education/ (Accessed: 5 June 2019).

Crew, N. (2018) interviewed by the author, 19 April 2018.

Department for Business, Innovation and Skills. (2009). *Skills for growth:*

The national skills strategy Cm 7641. Norwich: TSO (The Stationery Office).

Department for Business, Innovation and Skills (2013). Professor John Perkins' Review of Engineering Skills. London: DBIS. Available at: https://assets.publishing.service.gov.uk/government/uploads/system/uploads/attachment_data/file/254885/bis-13-1269-professor-john-perkins-review-of-engineering-skills.pdf (Accessed: 30 June 2020).

Department for Education and Skills. (2001). *Schools: Achieving Success* Cm 5230. Norwich: HMSO.

Department for Education and Skills. (2003). *14-19: Opportunity and Excellence*. Annesley: DfES Publications.

Department for Education and Skills. (2004a). *14-19: Opportunity and Excellence – Progress Report*. Annesley: DfES Publications

Department for Education and Skills. (2004b). *14-19 Curriculum and Qualifications Reform: Final Report of the Working Group on 14-19 Reform* [the Tomlinson Report]. Annesley: DfES Publications.

Department for Education and Skills. (2006). *Academy programme reaches halfway mark* [press release]. 16 March. Available at: https://www.wired-gov.net/wg/wg-news-1.nsf/54e6de9e0c383719802572b9005141ed/bea5f1f2ea575eca802572ab004bcc25?OpenDocument (Accessed: 9 March 2020).

Department for Education. (2010). Michael Gove to the Edge Foundation. 9 September. Available at: https://www.gov.uk/government/speeches/michael-gove-to-the-edge-foundation (Accessed: 12 February 2018). Note: at the time of writing, the website stated incorrectly that the speech was delivered on 29 September 2011.

Department for Education. (2011). *Review of vocational education – the Wolf Report*. London: Department for Education.

Department for Education. (2011b). *Wolf Review of Vocational Education – Government Response*. London: Department for Education. Available at: https://www.gov.uk/government/uploads/system/uploads/attachment_data/file/180868/Wolf-Review-Response.pdf (Accessed: 12 February 2018).

Department for Education. (2011c). 'Performance table reform and transparency will raise standards and end perverse incentives'. 20 July. Available at: https://www.gov.uk/government/news/performance-table-reform-and-transparency-will-raise-standards-and-end-perverse-incentives (Accessed: 22 May 2020).

Department for Education. (2014). *School Performance Tables, 2013*. Available at: http://webarchive.nationalarchives.gov.

uk/20150901232935/http://www.education.gov.uk/cgi-bin/schools/performance/2013/school.pl?urn=136933&superview=sec (Accessed: 21 February 2018).

Department for Education. (2016a). *Key stage 5 - institution level tables (SFR05/2016)*. Available at: https://www.gov.uk/government/uploads/system/uploads/attachment_data/file/493156/SFR052016_KS5_Institution.xlsx (Accessed: 22 February 2018).

Department for Education (2016b). *Report Of The Independent Panel on Technical Education* [The Sainsbury Report]. London: Department for Education. Available at: https://assets.publishing.service.gov.uk/government/uploads/system/uploads/attachment_data/file/536046/Report_of_the_Independent_Panel_on_Technical_Education.pdf (Accessed: 29 May 2020).

Department for Education (2017). *Post-16 technical education reforms: T level action plan*. London: Department for Education. Available at: https://assets.publishing.service.gov.uk/government/uploads/system/uploads/attachment_data/file/760829/T_Level_action_plan_2017.pdf (Accessed: 29 May 2020).

Department for Education (2018). *Digital and science engineering to be taught at new college* [press release]. 18 June. Available at: https://www.gov.uk/government/news/digital-and-science-engineering-to-be-taught-at-new-college (Accessed: 2 June 2020).

Department for Education (2019). *Independent Panel Report to the Review of Post-18 Education and Funding* (CP 117) (the Augar Report). Norwich: HMSO. Available at: https://assets.publishing.service.gov.uk/government/uploads/system/uploads/attachment_data/file/805127/Review_of_post_18_education_and_funding.pdf (Accessed: 1 July 2020).

Department for Education (2020). *School performance tables 2019*. Available at: https://www.compare-school-performance.service.gov.uk/find-a-school-in-england (Accessed: 22 April 2020).

Department for Education data (2019a). *Education and Training by Sector Subject Area – Number of Enrolled Aims* [online]. Available at: https://assets.publishing.service.gov.uk/government/uploads/system/uploads/attachment_data/file/872322/Education-and-training-aim-participation-SSA-gender_1516-1819.xls (Accessed: 1 July 2020).

Department for Education data (2019b). *Apprenticeships by Sector Subject Area – Number of Participants* [online]. Available at: https://assets.publishing.service.gov.uk/government/uploads/system/uploads/attachment_data/file/848363/

Apprenticeship_Participation_1415_1819_final_v0.2.xlsx (Accessed: 1 July 2020).

Department for Education data (2019c). 16 to 18 destination measures 2017 to 2018 [online]. Available at: https://www.gov.uk/government/statistics/destinations-of-ks4-and-16-to-18-ks5-students-2018 (Accessed: 13 July 2020).

Dixon, L. (2011). 'JCB engineers the opening of specialist academy', *The Times*. 7 January. Available at: https://www.thetimes.co.uk/article/jcb-engineers-the-opening-of-specialist-academy-v3zskxzr6jw. (Accessed: 22 May 2020).

Doyle, M. (2018) interviewed by the author, 16 October 2018.

Drysdale, L. (2019). 'This is what Sheffield student has in store on International Space University course'. *Yorkshire Post*. 25 June. Available at: https://www.yorkshirepost.co.uk/news/what-sheffield-student-has-store-international-space-university-course-1753937 (Accessed: 3 June 2020).

Duke of York (2013). Transcript of a speech by HRH The Duke of York, 8 July 2013. [Unpublished item].

Duncan, E. (2016). 'The march of the makers must start in class'. *The Times*, 13 August. Available at: https://www.thetimes.co.uk/article/the-march-of-the-makers-must-start-in-class-698wm2dqn (Accessed: 13 February 2019).

Edwards, R. (1960). *The Secondary Technical High School*. London: University of London Press.

Engineering UK (2019). *Key Facts & Figures: Highlights from the 2019 Update to the Engineering UK Report*. London: Engineering UK. Available at: https://www.engineeringuk.com/media/156186/key-facts-figures-2019.pdf (Accessed: 30 June 2020).

Engineering UTC North East Lincolnshire (2019). *GCSE & A Level Results 2019* [online]. 22 August. Available at: https://www.enlutc.co.uk/results-2019/ (Accessed: 21 February 2020).

Fowler, G. (2018) interviewed by the author, 25 July 2018.

Garner, R. (2013). 'Prince Andrew: "We must give young people confidence in the workplace"', *The Independent*. 8 December. Available at: http://www.independent.co.uk/news/education/education-news/prince-andrew-we-must-give-young-people-confidence-in-the-workplace-8991823.html (Accessed 7 March 2018).

Garner, R. (2016). 'The serious matter of role-playing doctors and nurses'. *The Independent*, 5 February.

Gove, M. (2010). 'All pupils will learn our island story' [online].

Conservative Party. Available at: https://conservative-speeches.sayit.mysociety.org/speech/601441 (Accessed: 3 June 2020).

Gove, M. (2017). 'My lesson from the latest schools scandal'. *The Times*, 10 February.

Government Communications Headquarters (2018). Letter from Head of Station GCHQ Scarborough to Charles Parker, Baker Dearing Educational Trust, 29 March. [Unpublished item].

Gray J., (2010). 'It's all a bit technical', *The Manufacturer*, 8 December. Available at: https://www.themanufacturer.com/articles/its-all-a-bit-technical/ (Accessed: 16 October 2017).

Green, M. (2016). 'Kenneth Baker and his guerrilla war over skills'. *Financial Times*, 1 May. Available at: https://www.ft.com/content/33044938-0de8-11e6-ad80-67655613c2d6 (Accessed: 13 February 2019).

Greening, J. (2017). Letter from Justine Greening to Lord Baker, 4 March. [Unpublished item].

Halliday, M. (2018) interviewed by the author, 26 April 2018.

Halstead A. (2011). Transcript of a speech by Prof Alison Halstead, 1 November 2011. [Unpublished item].

Harbourne, D. (2013). *Report to BDT Curriculum Working Group*, November 2013. [Unpublished item].

Harbourne, D. (2014). *Case study: Liverpool Life Sciences University Technical College*. London: Edge Foundation.

Harbourne, D. (2018a). Transcript of points made during a discussion with students and former students at UTC Sheffield, 19 April.

Harbourne, D. (2018b). Transcript of points made during a discussion with students at UTC Reading, 26 April.

Harper, J. (2018) interviewed by the author, 22 February 2018.

Harrison, M. (2011). *Respected: Technical Qualifications Selected for use in University Technical Colleges*. London: Edge Foundation. Available at: https://www.raeng.org.uk/RAE/media/General/News/Documents/Respected-Matthew-Harrison.pdf (Accessed 22 February 2018).

Hayes, J. (2011). Transcript of a speech by The Minister of State for Skills and Lifelong Learning, John Hayes. 1 November. [Unpublished item.]

Hilton C., (2011). Transcript of a speech by Chris Hilton, 1 November. [Unpublished item].

HC 87 (16 March 2020). *University Technical Colleges*. Evidence given by Jonathan Slater, Permanent Secretary, Department for Education to the Public Accounts Committee. Available at: https://committees.

parliament.uk/work/37/university-technical-colleges/publications/. (Accessed: 2 June 2020).

HC 87 (June 2020). *University Technical Colleges. House of Commons Committee of Public Accounts Fifth Report of Session 2019-21.* Available at: https://committees.parliament.uk/publications/1371/documents/12585/default/ (Accessed: 10 June 2020).

HC 691 (18 April 2018). *Delivering STEM skills for the economy.* Evidence given by Jonathan Slater, Permanent Secretary, Department for Education to the Public Accounts Committee. Available at: http://data.parliament.uk/writtenevidence/committeeevidence.svc/evidencedocument/public-accounts-committee/delivering-stem-skills-for-the-economy/oral/81686.html (Accessed: 29 May 2020).

HC 859 (18 December 2013). *Responsibilities of the Secretary of State.* Evidence given by Rt Hon Michael Gove MP, Secretary of State for Education to the House of Commons Education Committee – uncorrected transcript. Available at: https://www.parliament.uk/documents/commons-committees/Education/EdC181213.pdf (Accessed: 5 April 2019).

HC Deb (23 March 2011) vol. 525, col. 960.

HC Deb (10 October 2011) vol. 533, col. 63.

HC Deb (21 June 2012) vol. 546, col. 1025.

Higher Education Statistics Agency (2020). *HE Student Enrolments 2014/15 to 2018/19* [online]. Available at: https://www.hesa.ac.uk/data-and-analysis/students/what-study (Accessed: 1 July 2020).

Hinds, D. (2018). *Damian Hinds Technical Education Speech* [online]. Available at: https://www.gov.uk/government/speeches/damian-hinds-technical-education-speech (Accessed: 22 June 2020).

HL Deb (21 June 2006) vol. 477, col. 863.

HL Deb (13 December 2006) vol. 687, col. 1557.

HL Deb (8 November 2007) vol. 696, col. 221.

HL Deb (8 December 2008) vol. 706, cols. 512-3.

HL Deb (8 December 2008) vol. 706, col. 553.

HL Deb (8 December 2008) vol. 706, col. 573.

HL Deb (8 December 2008) vol. 706, col. 580.

HL Deb (1 February 2017) vol. 778, col. 1220.

HL Deb (22 February 2017) vol. 779, cols. 53-70GC.

HM Treasury. (2006). *Prosperity For All In The Global Economy – World Class Skills* (the Leitch Report). Norwich: HMSO.

Holliday M. (2013). Transcript of a speech by Mike Holliday, 8 July 2013. [Unpublished item].

Howlett, J. (2019) interviewed by the author, 18 July.

Hurst, G. (2011). 'Selection at 14 will drive revolution in schooling', *The Times*, 7 January. Available at: https://www.thetimes.co.uk/article/selection-at-14-will-drive-revolution-in-schooling-lkzp82hnqpl (Accessed: 22 May 2020).

Joint Council for Qualifications (2010 and 2019). *Provisional GCSE results* (England only) [online]. Available at: https://www.jcq.org.uk/examination-results/ (Accessed: 26 June 2020).

Joint Council for Qualifications (2019). *Provisional GCE A Level Results - June 2019* (England Only) [online]. Available at: https://www.jcq.org.uk/examination-results/a-levels/2019/main-results-tables (Accessed: 29 May 2020).

Kettlewell, K., Bernardinelli, D., Hillary, J. and Sumner, C. (2017). *University Technical Colleges: Beneath the Headlines. NFER Contextual Analysis*. Slough: NFER.

Linford, N. (2013). 'Back off, Baker', *FE Week*, 29 November. Available at: https://feweek.co.uk/2013/11/29/baker-defends-utc-under-recruitment/ (Accessed: 20 March 2018).

Manchester Evening News (2011). '24 new tech colleges planned', 23 March 2011. Available at : https://www.manchestereveningnews.co.uk/business/business-news/24-new-tech-colleges-planned-856998 (Accessed: 22 May 2020).

Mann, A. and Virk, B. (2013). *Profound employer engagement in education: what it is and options for scaling it up*. London: Edge Foundation. Available at: https://www.educationandemployers.org/wp-content/uploads/2014/06/profound_employer_engagement_published_version.pdf (Accessed: 27 February 2019).

McCrone T., Martin K., Sims D. and Rush C. (2017). *Evaluation of University Technical Colleges report – year one*. Slough: NFER.

McCrone, T., White, R., Kettlewell, K., Sims, S. and Rush, C. (2019). *Evaluation of University Technical Colleges*. Slough: NFER.

McKinsey Global Institute (2017). *Jobs Lost, Jobs Gained: Workforce Transitions in a Time of Automation* [online]. Available at: https://www.mckinsey.com/featured-insights/future-of-work/jobs-lost-jobs-gained-what-the-future-of-work-will-mean-for-jobs-skills-and-wages (Accessed: 30 June 2020).

Mills D. (2013). Transcript of a speech by David Mills, 8 July 2013. [Unpublished item].

Mujtaba, T. and Reiss, M. (2013). 'What sort of girl wants to study physics after the age of 16? Findings from a large-scale UK survey'.

International Journal of Science Education [online]. Vol. 35, No. 17, pp. 2979-2998. Available at: https://doi.org/10.1080/09500693.2012.681076 (Accessed: 10 January 2019).

Musuumba M. (2013). Transcript of a speech by Mwaka Musuumba, 8 July 2013. [Unpublished item].

National Audit Office (2019). *Investigation into University Technical Colleges: Report by the Comptroller and Auditor General* [HC 101]. Available at: https://www.nao.org.uk/wp-content/uploads/2019/10/Investigation-into-university-technical-colleges.pdf (Accessed: 2 June 2020).

Newton, O. (2016). 'Future careers' [online]. *Fabian Society*, 22 July. Available at: https://fabians.org.uk/future-careers/ (Accessed: 27 March 2019).

Northern Schools Trust (2020). *Liverpool Life Sciences UTC* [online]. Available at: https://northernschoolstrust.co.uk/our-schools/ (Accessed 20 February 2020).

Nye, N. and Camden, B. (2014). 'Jobs for the boys at UTCs'. *Schools Week*. 10 October.

Office of National Statistics (2019). *Overeducation and hourly wages in the UK labour market* [online]. Available from: https://www.ons.gov.uk/economy/nationalaccounts/uksectoraccounts/compendium/economicreview/april2019/overeducationandhourlywagesintheuklabourmarket2006to2017 (Accessed: 30 June 2020).

Ofsted (2007). *The Key Stage 4 curriculum: Increased Flexibility and Work-Related Learning.* Manchester: Ofsted.

Ofsted (2011a). *Meeting Technological Challenges? Design and Technology in Schools 2007–10.* Manchester: Ofsted. Available at: https://assets.publishing.service.gov.uk/government/uploads/system/uploads/attachment_data/file/413705/Meeting_technological_challenges.pdf (Accessed: 10 January 2019)

Ofsted (2011b). *Girls' Career Aspirations.* Manchester: Ofsted. Available at: https://assets.publishing.service.gov.uk/government/uploads/system/uploads/attachment_data/file/413603/Girls__career_aspirations.pdf (Accessed: 10 January 2019).

Ofsted (2013). I*nspection report: Black Country UTC, 29-30 January 2013*. Manchester: Ofsted. Available at: https://files.ofsted.gov.uk/v1/file/2191073 (Accessed: 21 February 2018).

Ofsted (2014a). *Inspection report: Hackney UTC, 15-16 January 2014.*

Manchester: Ofsted. Available at: https://files.ofsted.gov.uk/v1/file/2322863 (Accessed 6 March 2018).

Ofsted (2014b): *Inspection report: Central Bedfordshire UTC, 18-19 March 2014.* Manchester: Ofsted. Available at: https://files.ofsted.gov.uk/v1/file/2388961 (Accessed: 2 March 2018).

Ofsted (2014c). *Monitoring inspection visit to Hackney University Technical College, 9 May 2014.* Manchester: Ofsted. Available at: https://files.api.ofsted.gov.uk/v1/file/2391265 (Accessed: 6 March 2018).

Ofsted (2014d). *Inspection report: Aston University Engineering Academy, 4-5 June 2014.* Manchester: Ofsted. Available at: https://files.ofsted.gov.uk/v1/file/2405690 (Accessed: 6 March 2018).

Ofsted (2015a). *Inspection report: Black Country UTC, 10-11 March 2015.* Manchester: Ofsted. Available at: https://files.ofsted.gov.uk/v1/file/2480479 (Accessed: 8 March 2018).

Ofsted (2015b). *Inspection report: UTC Reading 19-20 May.* Manchester: Ofsted. Available at: https://files.ofsted.gov.uk/v1/file/2488411 (Accessed: 3 June 2020).

Ofsted (2016b). *Inspection report: UTC Sheffield 2-3 February 2016.* Manchester: Ofsted. Available at: https://files.api.ofsted.gov.uk/v1/file/2546631 (Accessed: 21 February 2019).

Ofsted (2017). *Inspection report: Lincoln UTC, 21-22 February.* Manchester: Ofsted. Available at: https://files.ofsted.gov.uk/v1/file/2667985 (Accessed: 3 June 2020).

Ofsted (2018a). *Inspection report: Engineering UTC Northern Lincolnshire 6-7 February 2018.* Manchester: Ofsted. Available at: https://files.api.ofsted.gov.uk/v1/file/2760427 (Accessed: 5 June 2019).

Ofsted (2018b). *Short inspection of The JCB Academy 25 September 2018.* Manchester: Ofsted. Available at: https://files.api.ofsted.gov.uk/v1/file/50036745 (Accessed: 3 April 2019).

Ofsted (2018c). *Inspection report: Bristol Technology and Engineering Academy 16-17 October.* Manchester: Ofsted. Available at: https://files.ofsted.gov.uk/v1/file/50043765 (Accessed: 3 June 2020).

Ofsted (2018d). *Inspection report: London Design and Engineering UTC 16-17 October.* Manchester: Ofsted. Available at: https://files.ofsted.gov.uk/v1/file/50040492 (Accessed: 3 June 2020).

Ofsted (2019a). *Inspection report: Greater Peterborough UTC 12-13 February.* Manchester: Ofsted. Available at: https://files.ofsted.gov.uk/v1/file/50063651 (Accessed: 3 June 2020).

Ofsted (2019b). *Inspection report: Bolton UTC 13-14 March.* Manchester:

Ofsted. Available at: https://files.ofsted.gov.uk/v1/file/50068032 (Accessed: 3 June 2020).

Ofsted (2019c). *Inspection report: UTC Leeds 26-27 March*. Manchester: Ofsted. Available at: https://files.ofsted.gov.uk/v1/file/50076133 (Accessed: 3 June 2020).

Ofsted (2019d). *Inspection report: Energy Coast UTC 5-6 June*. Manchester: Ofsted. Available at: https://files.ofsted.gov.uk/v1/file/50091840 (Accessed: 3 june 2020).

Ofsted (2019e). *Inspection report: Elstree UTC 24-25 September*. Manchester: Ofsted. Available at: https://files.ofsted.gov.uk/v1/file/50117557 (Accessed: 24 April 2020).

Ofsted (2020a): *Inspection report: Liverpool Life Sciences UTC, 21-22 January*. Manchester: Ofsted. Available at: https://files.ofsted.gov.uk/v1/file/50145909 (Accessed 24 April 2020).

Ofsted (2020b). *Inspection report: UTC Sheffield City Centre, 12-13 February*. Manchester: Ofsted. Available at: https://files.ofsted.gov.uk/v1/file/50149110 (Accessed: 3 June 2020).

Ofsted (2020c). *Inspection report: Health Futures UTC 25-26 February*. Manchester: Ofsted. Available at: https://files.ofsted.gov.uk/v1/file/50149687 (Accessed: 3 June 2020).

Ofsted (2020d). *Inspection report: Ron Dearing UTC 11-12 March 2020*. Manchester: Ofsted. Available at: https://files.ofsted.gov.uk/v1/file/50151274 (Accessed: 10 June 2020).

Parker, C. (2015). 'Parents have clear views on the education system, it's time they were heard', *Daily Telegraph*. 29 October. Available at: https://www.telegraph.co.uk/education/educationopinion/11960332/Parents-have-clear-views-on-the-education-system-its-time-they-were-heard.html (Accessed: 27 May 2020).

Parker, C. (2017). Presentation to UTC governors, July. [Unpublished item].

Parker, C. (2018) interviewed by the author, 14 February 2018.

Pashley, S. (2018). Presentation to UTC annual conference, 17 July. [Unpublished item].

Paterson, S. (2017). 'Justine Greening: Scarborough UTC is just what young people need'. *Scarborough News*, 5 April. Available at: https://www.thescarboroughnews.co.uk/news/justine-greening-scarborough-utc-is-just-what-young-people-need-1-8343628 (Accessed: 5 April 2019),

Pilsworth-Straw, E. (2019) interviewed by the author, 5 July 2019.

Revington, T. (2014). *Learning the Lessons from the Early Years of*

Establishing University Technical Colleges. London: Baker Dearing Educational Trust. [Unpublished item].

Reynolds, R. (2018) interviewed by the author, 19 April 2018.

Robertson, A. (2016). 'Numbers falling, closing down – University Technical College revolution fails to deliver'. *FE Week*, 8 February. Available at: https://feweek.co.uk/2016/02/08/numbers-falling-closing-down-university-technology-college-revolution-fails-to-deliver/ (Accessed: 12 February 2019).

Robertson, A. (2018). 'Pupil destinations will trump Progress 8 as headline measure for UTCs'. *FE Week*, 21 August 2018. Available at: https://schoolsweek.co.uk/pupil-destinations-will-supersede-progress-8-as-headline-measure-for-utcs-and-studio-schools/ (Accessed: 29 May 2020).

Rosen, W. (2010). *The Most Powerful Idea in the World: A Story of Steam, Industry, and Invention*. London: Random House.

Royal Navy (2017): *Young Engineers Take On Disaster Relief Challenge for British Science Week* [online]. 17 March. Available at: https://www.royalnavy.mod.uk/news-and-latest-activity/news/2017/march/17/170317-young-engineers-take-on-disaster-relief-challenge-for-british-science-week (Accessed: 27 March 2019).

Sainsbury, D. (2010). *Report from Lord Sainsbury to Lord Mandelson on the Professional Registration of Technicians*, April 2010. Available at: http://www.gatsby.org.uk/uploads/education/reports/pdf/13-lord-sainsbury-report-on-technicians-april-2010.pdf (Accessed 27 February 2018).

Sainsbury, D. (2011). 'Welcome and introduction' in *Technical Education for the 21st Century*. London: The Gatsby Charitable Foundation. Available at: https://www.gatsby.org.uk/uploads/education/reports/pdf/7-technician-conference-report.pdf (Accessed 27 February 2018).

Sainsbury, D. (2012). Letter to Lord Baker. 10 July. [Unpublished item].

Satchwell, K. and Maher, V. (2015). Report to Baker Dearing Educational Trust board of trustees. November. [Unpublished item].

Schools Adjudicator (2013). *Determination of Case Reference ADA2424 (The JCB Academy)* [online]. Available at: https://assets.publishing.service.gov.uk/government/uploads/system/uploads/attachment_data/file/295558/ada2424_jcb_academy_staffordshire_29august13.pdf (Accessed: 10 January 2019).

Schuhmacher, M. (2015). *UTCs Delivering Apprenticeships*. [Unpublished item].

Sedgmore, L. (2014). 'Are UTCs another "purely political vanity project" set for the history books?' *FE Week*. 11 July. Available at: https://feweek.co.uk/2014/07/11/are-utcs-another-purely-political-vanity-project-set-for-the-history-books/ (Accessed: 20 March 2018).

Smithers, A. (2011). 'It's the right age: pupils will know their strengths', *The Times*. 7 January. Available at: https://www.thetimes.co.uk/article/its-the-right-age-pupils-will-know-their-strengths-kn738znghph. (Accessed 22 May 2020).

Stark, M. (2015). *Viability of University Technical Colleges at KS4*. London: Baker Dearing Educational Trust. [Unpublished item].

Symonds J. (2013). Transcript of a speech by Joe Symonds, 8 July 2013. [Unpublished item].

Tabor, A. (2013). Message in support of proposals for the Global Academy. [Unpublished item].

Thacker, A. (2018), interviewed by the author, 26 April 2018.

The JCB Academy (2013). *JCB Engineering Challenges 2014/15: Overview*. Rocester: The JCB Academy. [Unpublished item].

The JCB Academy (2015). *UTCs and Apprenticeship Programmes: A Case Study*. [Unpublished item].

The JCB Academy (2019). Business Pathway [online]. Available at: https://jcbacademy-sixthform.com/business-pathway/ (Accessed: 3 April 2019).

The JCB Academy Trust (2006). *Charitable Objects*. Available at: https://apps.charitycommission.gov.uk/Showcharity/RegisterOfCharities/RemovedCharityMain.aspx?RegisteredCharityNumber=1120673&SubsidiaryNumber=0 (Accessed: 12 January 2018).

The Royal Family (2011). *The Prince of Wales and The Duchess of Cornwall Open the new JCB Academy in Rocester, UK* [online]. YouTube. [Viewed 9 February 2018]. Available at: https://www.youtube.com/watch?v=01l4mvwMHsE

The Studio Liverpool (2018). S*tudio School Students Create a Changing World for the Better* [online prospectus]. Available at: http://thestudioliverpool.uk/wp-content/uploads/2019/10/Studio-School-prospectus-2018.pdf (Accessed 20 February 2020).

Tomlinson M. (2017) interviewed by the author, 22 November 2017.

UTC Oxfordshire (2016). *Secretary of State for Education Visits UTC Oxfordshire* [online]. Available at: http://www.utcoxfordshire.org.uk/secretary-of-state-for-education-visits-utc-oxfordshire/ (Accessed: 13 February 2019).

UTC Reading (2013). Brief given to students at the start of a post-16 project delivered in 2013-14. [Unpublished item].

UTC Sheffield (2019). *Student experience* [online]. Available at: https://www.utcsheffield.org.uk/student-experience/ (Accessed 27 May 2020).

Vaughan, R. (2011). 'Osborne signals technical college surge', *Times Educational Supplement*. 25 March.

Wade, J. (2018) interviewed by the author, 23 January 2018.

Ward, N. (2017) interviewed by the author, 14 December 2017.

Ware, J. (2011). *The JCB Academy: Year One*. London: Baker Dearing Educational Trust.

Weale, S. (2017). '£9m Greater Manchester college closes after three years due to lack of pupils'. *The Guardian*, 7 February. Available at: https://www.theguardian.com/education/2017/feb/07/greater-manchester-university-technical-college-closes-three-years (Accessed: 29 May 2020).

Wilby, P. (2016). 'Rich List landlords, apologies for Hillsborough, and Lord Baker – my unlikely Tory comrade'. *New Statesman*. 19 April. Available at: https://www.newstatesman.com/politics/sport/2016/04/surely-its-not-too-late-sun-sack-kelvin-mackenzie (Accessed 19 February 2019).

Wilson, J. (2011). 'Young Apprenticeships (YA) - No New Starts'. Note sent by the Young People's Learning Agency to local authorities. 28 March. [Unpublished item].

Wilson, R. (2018) interviewed by the author, 26 April 2018.

WISE Campaign (2019). *2019 Workforce Statistics – One Million Women in STEM in the UK* [online]. Available at: https://www.wisecampaign.org.uk/statistics/2019-workforce-statistics-one-million-women-in-stem-in-the-uk/ (Accessed: 29 May 2020).

Worsey, L. (2019) interviewed by the author, 30 August 2019.

Worth, J., Lynch, S., Hillary, J., Rennie, C. and Andrade, J. (2018). *Teacher Workforce Dynamics in England*. Slough: NFER. Available at: https://www.nfer.ac.uk/media/3111/teacher_workforce_dynamics_in_england_final_report.pdf (Accessed: 3 June 2020).

Wright, R. (2018) interviewed by the author, 19 April 2018.

INDEX

2Bio 95

Able UK 229
Activate Learning Education Trust 103, 104, 208
Adonis, Lord (Andrew) 9, 10, 29, 30, 36, 40, 45, 52, 74, 90, 217, 257, 262, 263
Advanced Manufacturing Research Centre (AMRC) 138
Agnew, Lord 243, 244, 250
Air Products 200
AldridgeUTC@MediaCityUK 282, 285
Allcooper 195
Alpha Manufacturing 210
Apps for Good 195
Arkwright, Richard 10
Ashton, Geoff 129
Aston University 57, 58, 59, 60
Aston University Engineering Academy 1, 44, 46, 49, 52, 57, 59, 60, 61, 72, 118, 132, 152, 198, 206, 207, 208, 282, 285
Atkins 45, 226
Augar, Philip 273

BAE Systems 239
Bailey, Michael 209, 210, 211
Baker Award 88, 89
Baker Dearing Educational Trust 1, 2, 17, 44, 45, 46, 49, 52, 56, 61, 72, 75, 78, 79, 80, 83, 84, 87, 88, 100, 110, 111, 116, 117, 119, 120, 121, 123, 124, 125, 126, 129, 130, 131, 132, 133, 134, 153, 154, 156, 161, 162, 166, 167, 168, 171, 172, 173, 193, 196, 197, 202, 208, 209, 219, 220, 221, 222, 223, 224, 225, 226, 227, 228, 230, 236, 237, 238, 239, 241, 242, 243, 244, 245, 246, 247, 250, 251, 252, 253, 254, 255, 256, 257, 258, 263, 265, 266, 267
Baker, Lord (Kenneth) 1, 2, 28, 29, 34, 35, 36, 37, 38, 39, 40, 41, 42, 43, 44, 45, 46, 47, 48, 49, 50, 51, 52, 53, 54, 57, 59, 75, 76, 80, 84, 91, 99, 110, 111, 112, 123, 124, 161, 162, 163, 164, 169, 170, 174, 193, 206, 215, 216, 217, 218, 219, 255, 257, 258, 260, 261

Bamford, Joseph Cyril 8
Bamford, Lord (Anthony) 8, 9, 74
Barclays Bank 231
Barker, Clive 256
Basi, Amarjit 46, 61, 62
Bather, Tom 88
Bedford College 115, 116
Bell, Charles 268
Bell, David 8, 9, 11, 100, 113, 177, 178, 179, 190
Bentley 6, 7, 10, 87, 179
Best, Fay 153, 156, 157, 158, 159
Bhattacharyya, Lord 173, 174
Birkenhead UTC 77
Birmingham City Council 57, 58
Birmingham City University 78, 79
Birmingham Science Park 59
Birtwistle, Gordon 45
Black Country UTC 1, 46, 52, 57, 61, 62, 63, 64, 66, 68, 69, 70, 72, 74, 84, 86, 87, 88, 114, 116, 117, 118, 119, 120, 122, 124, 152, 237, 264
Blair, Tony 10, 11, 29, 31, 34, 262, 275
Blower, Christine 54
Blunkett, David 30
Bmat Stem Academy 282, 286
BMW 60
Boden, Carly 231
Boeing 136
Boles, Nick 132, 133, 134, 161, 169
Bombardier 10, 16
Bosch 187, 190
Bosch Rexroth 179, 180
Boyle, Miss 6
BP 140
Bradford University 92
Bragg, Valerie 78, 79
Bright Futures Educational Trust 161
Bristol and South Gloucestershire UTC 73
Bristol Technology and Engineering Academy 118, 122, 152, 222, 246, 282, 286
Bri-Stor Systems 210
British Aerospace 115
British Army 105
British Steel 231, 234
Brooks, Gavin 107, 108
BT 114
Buckinghamshire UTC 73, 118, 129, 152, 156, 282, 286
Buckwell, Mr 259
Burnley UTC 73
Burton College 9
Butler, R.A. 37, 40, 43, 76
Byrne, John 194, 195

Cadbury 60
Cambridge Academy for Science and Technology 282, 286
Cambridge University 14, 23, 25, 59, 137, 259
Cameron, David 39, 40, 169, 263
Capper, Stephen 54
Caterpillar 64
Cavendish, Camilla 133
Central Bedfordshire College 115
Central Bedfordshire Council 115
Central Bedfordshire UTC 114, 115, 116, 122, 124, 152, 161, 237
Centrica 229
Chamberlin plc 64, 65
Charity Commission 12, 44
Church of England 35
Cisco 98, 109
City and Guilds of London 58, 166
City Technology Colleges 1, 28, 44, 221, 255, 261, 262
Clarke, Lewis 3
Clarke, Mr 24

Clugston 229
Collins Education 90
Compton Hospice 195
Connell, Simon 88, 89, 167, 168, 224, 225, 245, 257, 258, 259
Conservative Party 39, 41, 42, 47, 52, 132, 263
Continental Engineering Services 179, 210
Cook, Allan 45
Cornforth, Ken 246, 254, 257
Cornwall, Duchess of 17
Council for National Academic Awards (CNAA) 35
Craftsman Cup 104, 106
Cranfield University 115
Crewe Engineering and Design UTC 282, 286
Crew, Nick 136, 138, 140
Cristal 229, 231
Croda 94
Cropley, Andrew 135, 137
Cummings, Dominic 111

Daily Mail, The 111
Daily Telegraph, The 129, 162
Daventry UTC 73, 118, 120, 152, 237
Davies, Jill 95
Dearing, Lord (Ron) 1, 28, 34, 35, 36, 37, 40, 41, 43, 44, 45, 48, 51, 52, 57, 59, 200, 206, 262, 263, 275, 283, 285
Department for Business, Innovation and Skills 42, 43, 270
Department for Children, Schools and Families 42, 44, 58, 59
Department for Education and Skills (DfES) 9, 10, 11, 12, 13, 30, 31, 32, 48, 58
Department for Education (DfE) 9, 11, 30, 31, 32, 47, 48, 49, 50, 51, 52, 55, 58, 72, 75, 99, 100, 101, 110, 113, 119, 121, 123, 125, 126, 131, 133, 136, 152, 162, 164, 166, 167, 170, 171, 172, 216, 220, 221, 223, 224, 228, 230, 236, 239, 241, 242, 243, 244, 246, 247, 248, 249, 250, 251, 254, 255, 256, 258, 265, 267, 272, 273, 274, 275, 282, 286
Department for Trade and Industry 34
Department of Health 110
DePuy Synthes 26, 27
Derby Manufacturing UTC 160, 186, 187, 194, 208, 251, 282, 286
Discovery School 126
Dixon, Laura 52
Djanogly, Harry 29
Doncaster UTC 249, 250, 256
Dove Engineering Centre 210
Doyle, Marc 230, 231, 232, 233, 234, 235, 254, 257
Drysdale, Laura 277
Duke of Edinburgh Award 16
Duke of York Award for Technical Education 78, 80, 84
Dulverton Trust 256
Duncan, Emma 169
Dundee University 92
Durham University 59, 257, 284, 285

East London UTC 77
East Midlands Airport 5
East Midlands Helicopters 5
Edge, Callum 88
Edge Foundation 1, 2, 37, 45, 46, 47, 49, 50, 56, 82, 93, 163, 175, 198, 203, 204, 256
Education Act, 1944 1, 40, 260
Education and Employers Taskforce 203

Education and Skills Funding Agency (ESFA) 116, 212, 224, 230, 256
Education Funding Agency 75, 130
Edwards, Reese 1
Ellison, Bradley 139
Elstree UTC 77, 118, 152, 154, 156, 252, 283, 285
Elutec 77, 118, 282, 286
Emsi 225
Energy Coast UTC 77, 118, 248, 251, 282, 285
Engineering Construction Industry Training Board 229
Engineering Council 80, 82, 149
Engineering Learning Mentors (ELMs) 18, 19
Engineering UK 271
Engineering UTC Northern Lincolnshire 229, 230, 232, 233, 234, 235, 248, 254, 283, 286
Equal Opportunities Commission 149
European Space Agency 277, 281

F1 60, 141, 143, 144, 145, 146, 155
F1 in Schools 60
Ferrari 5
FE Week 121, 160
Field, Frank 45, 75
Financial Times 163
Fink, Stanley 90
Finning 65
Firth Rixson 136
Flamingo Land 195
Fogarty, Kathy 45
Fowler, Geoffrey 199
Freeman, Ashley 198
Fujitsu 155, 226

Garfield Weston Foundation 256
Garner, Richard 160

Gatsby Charitable Foundation 45, 80, 82, 83, 129, 256
GCHQ 196, 197
GL Assessment 171, 256
Global Academy UTC 156, 197, 283, 285
Global Radio 197
Gothenburg Technical Gymnasium 9
Gove, Michael 40, 41, 42, 47, 48, 49, 50, 53, 64, 73, 76, 98, 110, 111, 112, 122, 123, 124, 173, 174, 263, 268, 269
Greater Manchester Sustainable Engineering UTC 118, 172, 173, 174, 237
Greater Peterborough UTC 222, 283, 285
Greene, Mr 6, 24
Greening, Justine 169, 170, 215, 216, 219, 220
Green, Miranda 163
Guardian, The 172
Gwinnet, Paula 19, 24, 279

Hackney Community College 80, 114
Hackney UTC 114, 115, 117, 118, 122, 124, 152, 237
Halfon, Robert 75, 169
Halliday, Michael 103, 104
Halstead, Alison 46, 57, 59, 60
Hammond, Philip 169
Harlow UTC 77
Harper Adams University 4, 10, 13, 14, 19, 22, 179, 182, 183, 184, 188, 189
Harper, Joanne 101, 102, 168
Harrington, Dominic 246
Harrison, Matthew 15, 56, 113
Haughton Design 62, 65, 69, 70, 195
Hayes, Alex 129
Hayes, John 49, 76

Hayter, Paul 139
Health Futures UTC 156, 160, 197, 253, 283, 285
Heathrow Aviation Engineering UTC 77, 118, 156, 286
Hennessy, Peter 170
Hewlett Packard 226
Hilton, Chris 61, 62, 64, 117
Hilton, Steve 39, 53
Hinds, Damian 249
HM Treasury 12, 34, 110, 131, 167, 236
Holliday, Brian 70, 71
Homerton University Teaching Hospital 114
Hornby, David 93, 95
Houghton Regis 73
House of Commons 29, 73, 74, 122, 170, 228, 237, 238
House of Lords 8, 10, 29, 34, 35, 36, 40, 44, 216, 217, 218, 219, 262, 263
Howlett, Jodie 277, 278, 279, 280, 281
HSBC 136
Humber Renewables and Engineering UTC 160, 229
Hunt, Sally 54

Independent, The 88, 112, 160
Industrial Revolution 10, 164
Industrial Strategy Council 272, 273
Inspec Solutions 139
Inspiration Academy 257
Institute for Apprentices 212
Institute for Apprenticeships and Technical Education 166
Institute of Education 46
Institution of Engineering and Technology 179, 226
International Space University (ISU) 277
IPPR 243

Jaguar Land Rover 6, 88
JCB 3, 4, 6, 8, 10, 11, 16, 17, 74, 179, 180, 181, 183, 186, 187, 190, 191, 194, 207, 210, 212, 213, 277
JCB Academy 1, 3, 4, 5, 6, 8, 10, 11, 12, 13, 14, 16, 17, 19, 21, 22, 23, 24, 25, 26, 27, 37, 46, 48, 52, 54, 72, 74, 78, 84, 86, 87, 88, 113, 118, 122, 131, 132, 151, 152, 161, 168, 177, 178, 179, 180, 181, 182, 183, 184, 185, 186, 187, 188, 189, 190, 192, 194, 195, 207, 208, 209, 210, 212, 213, 214, 221, 254, 264, 277, 278, 279, 280, 281, 283, 285
JCB Aviation 5
John Leggott College 232
John Moores University 92
Johnson and Johnson 26
Joint Council for Qualifications 151, 269

Kalms, Stanley 29
Kelly, Ruth 11
Kier 155
King's College London 50
Kingshurst Achievement Award 78, 79
Kingsmead Technology School 64
Knight, Jim 40
Kokkolas, Lena 45
Kraft 136

Labour Party 9, 29, 35, 37, 39, 41, 45, 47, 75, 100, 218, 263
Lancashire 129, 237
Land, David 257
Lavender International 142
Leeds Local Enterprise Partnership 231
Leeds UTC 157, 158, 222, 257
Lego 155
Leigh Academies Trust 161

Leigh UTC 118, 233, 244, 245, 248, 257, 283, 285
Leitch Report 9, 10, 11, 12
Letwin, Oliver 39
Liberal Democrats 47, 217, 218, 263
Lincoln UTC 77, 118, 163, 221, 283, 286
Linford, Nick 121
Liverpool Community Health 95
Liverpool Life Sciences UTC 73, 90, 93, 94, 95, 118, 152, 154, 156, 197, 252, 283, 285
Liverpool Low Carbon and SuperPort UTC 77
Liverpool University 92, 95
Lloyd, Phil 94
Local Improvement Finance Trusts 110
London Design and Engineering UTC 199, 222, 248, 258, 283, 285
London Development Agency 114

Madeley Academy 132
Maher, Vic 131, 132
Manchester Evening News 54
Mandelson, Lord 82
Man Group plc 90
May, Theresa 169
McKenzie, Lord 41
McKinsey & Company 125
McKinsey Global Institute 272
Medway UTC 160, 258
Mercers' Company 29, 30, 45
Michael Bishop Foundation 256
Michelin 210
Microsoft 98, 102
Midcounties Co-operative Pharmacy 160
Mills, David 69, 70
Ministry of Power 34
Mitchell, Peter 9, 37, 45
Morant, Robert 260, 268

Morgan Tucker 195
Morris, Estelle 30, 31, 217, 218
Morrisons 191
Mott MacDonald 145
Mulberry UTC 156, 248, 283, 286
Musuumba, Mwaka 68, 69

Nash, Lord 123, 124, 140, 215, 216, 219
National Apprenticeship Service 206
National Audit Office (NAO) 237, 238, 239, 240
National Foundation for Educational Research 2, 105, 174, 175, 176, 204, 205
National Grid 4, 59, 60, 64, 179
National Health Service (NHS) 114, 197
National Union of Teachers 54
Network Rail 4, 5, 10, 16, 98, 109, 180
Newcastle UTC 73
Newton, Olly 198, 199
Nicholls, Lee 99
Nintendo 91
Norcup, Miss 25
Norfolk UTC 77, 285
Norris, Keith 9
North East Futures UTC 286
Northern General Hospital 143
Northern Schools Trust 90, 93, 95
North Lincolnshire Council 229, 232
North Lindsey College 229, 232
North Liverpool Academy 90, 92
North Liverpool Academy Trust 90
Nottingham University 26, 73, 126, 136
Nottingham UTC 73
Novartis 94, 95
Nuffield 23, 246

OCR (Oxford, Cambridge and RSA Examinations) 14, 15, 113, 120, 166

Office of National Statistics 272
Ofsted 33, 34, 60, 61, 66, 67, 90, 95, 96, 103, 114, 115, 116, 117, 119, 120, 121, 127, 138, 139, 150, 151, 161, 168, 171, 172, 181, 216, 221, 222, 223, 224, 234, 235, 245, 246, 247, 248, 251, 252, 253, 256, 262, 275
Ollis, Mr 6, 24, 25
OnePoll 225, 226
Orbital Gas Systems 210
Osborne, George 40, 53, 54, 76, 161, 169, 263, 264
Outwood Grange Academies Trust 229, 230
Oxford and Cherwell Valley College 99
Oxford University 14, 59, 137

Parker, Charles 45, 72, 73, 80, 123, 124, 125, 129, 130, 196, 215, 224, 242, 257, 258
Partnerships for Schools 45, 75
Pascoe, Sarah 257
Pashley, Sarah 200, 201
Pearson 113, 166
Pedroza, Anna 129, 225
Perkins, John 270
Peter Brett Associates 98, 103, 208
Peter Cundill Foundation 256
Phillips 66 229, 230, 231, 234
Pickering, Daniel 139
Pilsworth-Straw, Ella 19, 20, 21, 27
Post Office 35
Pritchard, Paul 9, 11, 12, 46
Proctor and Gamble 18
Pro Lab Diagnostics 95
Public Accounts Committee 228, 237, 238, 239, 240, 241
Puttnam, Lord 45
PWC 180, 190

RAF 145, 231
Reading Borough Council 99, 100
Reading College 99, 109
Reading University 99, 100, 107
Reading UTC 105, 108, 109, 122, 152
Redbourne 256
Revington, Tom 125, 126
Reynolds, Alex 66, 137, 138, 140, 147
Richardson, William 46
Rodillian Multi-Academy Trust 230
Rolls-Royce 4, 6, 10, 14, 16, 18, 20, 45, 59, 142, 143, 179, 187, 195, 208, 278
Ron Dearing UTC 200, 250, 253, 257, 258, 283, 285
Rose, John 44
Rosen, William 268
Royal Academy of Engineering 15, 56, 113, 153, 175, 179, 204, 270, 271, 281
Royal Electrical and Mechanical Engineers (REME) 104, 105, 107
Royal Family 17
Royal Greenwich UTC 118, 126, 129, 152, 154, 155, 161, 237, 255
Royal Liverpool and Broadgreen University NHS Trust 94
Royal Navy 18, 172, 191, 195, 196, 198, 239
Royal Shakespeare Company 94
Royal Society 94
RSA Examinations 14
RWDI Anemos 115

Sainsbury, Lord 80, 81, 82, 83, 164, 166, 273
Salesforce 103
Samuels, Jane 37, 57
Sandvik 64, 65
Satchwell, Kevin 131, 132

Sawyers Hall College 54
Scarborough News 215
Scarborough UTC 196, 197, 215, 283, 286
Schools Adjudicator 152
Schools Week 152, 153
Schuhmacher, Michael 208, 209
Science Council 80
Secondary Technical School, The 1
Sedgmore, Lynne 122
Semta Apprenticeship Service 206, 207
SFA 207, 208, 209
SGS Berkeley Green UTC 283, 286
Shakespeare, William 43, 261
Sheffield Chamber of Commerce 135
Sheffield College 135, 137, 138
Sheffield Hallam University 135, 144, 147, 153, 210, 213, 277, 280
Sheffield Sharks 142
Sheffield University 19, 92, 136
Sheffield UTC 122, 161
Siemens 18, 62, 64, 65, 68, 70, 95, 116, 145, 163, 264, 280
Siemens Gamesa Renewable Energy 200
Silverstone UTC 73, 118, 122, 152, 208, 283, 285
Simpkin, Theresa 257
Singleton Birch 229
Sir Charles Kao UTC 118
Sir Simon Milton Westminster UTC 283, 286
Skills Funding Agency 206, 207
Slater, Jonathan 170, 228, 239, 240, 241
Smithers, Alan 52
Smith & Nephew 201
Sneyd Community School 61, 64, 117
Sony 91
South Bank Engineering UTC 283, 286
South Devon UTC 160, 283, 286
South Oxfordshire UTC 160

South Staffordshire Water 62, 64, 65
Southwark UTC 73
South Wiltshire UTC 160, 237, 283
South Yorkshire Fire and Rescue Service 136
Spencer Group 200
Spooner, Mark 19
Stark, Michael 130, 131, 167
Storey, Lord 217, 218
Stratasys 64, 65
Stuart, Graham 122
Studio School 92, 93, 94, 95, 96, 97
Sullivan, Adam 226
Symonds, Joe 68, 87, 88

Tabor, Ashley 197
Tarmac 30
Tata 59, 136, 229
Teaching School Alliances 223
Technical and Further Education Act 216, 217, 220
Technician Council 82
Thacker, Andrew 107
Thames Water 103, 155, 199
The Inspiration Centre 233
Thomas Alleyne's High School 8, 9, 10, 17, 37
Thomas Linacre School 1
Thomas, Nigel 45, 80, 83
Thomas Telford School 30
Thomas Telford UTC 248, 283, 286
Thornton, Sullivan 230
Times Educational Supplement 54
Times, The 51, 52, 169, 173
Tomlinson Report 31, 32, 44, 45
Tomlinson, Sir Mike 30, 31, 32, 34, 44, 45, 80, 83, 84, 110, 257, 262, 273
Total 229
Tottenham UTC 118, 126, 237, 255

Toyota 10, 14, 16, 25, 88, 179, 180, 186, 187, 194, 195
Trentham Gardens 180
Tresham College 208
Turner, Georgia 88

ULTRA-PMES 180
Unilever 94, 95
University and College Union 54
University of Bedfordshire 115
University of Buckingham 2, 52
University of Central Lancashire 92
University of Exeter 46
University of Hull 229, 231
University of Liverpool 93
University of the Arts London 197
UTC Bluewater 77
UTC Bolton 156, 160, 222, 245, 250, 282, 286
UTC Cambridge 77, 118, 122
UTC Central Bedfordshire 119
UTC Harbourside 160, 237, 284
UTC Heathrow 103, 283
UTC Lancashire 118
UTC Leeds 283, 285
UTCMediaCityUK 160
UTC Network Trust 162, 220
UTC Norfolk 118, 122, 129, 283
UTC Oxfordshire 103, 170, 215, 250, 283, 285
UTC Plymouth 73, 119, 129, 152, 195, 242, 245, 283, 286
UTC Portsmouth 195, 284, 285
UTC Reading 73, 98, 100, 103, 106, 107, 119, 120, 155, 168, 173, 208, 248, 284, 285
UTC Sheffield 66, 73, 107, 119, 135, 136, 137, 138, 139, 140, 141, 144, 145, 146, 148, 152, 248, 252, 284, 285
UTC Sheffield Academy Trust 140
UTC Sheffield City Centre 66, 140, 141, 142, 143, 144, 145, 146, 147, 161, 163, 171, 172, 248, 252, 284, 285
UTC Sheffield Olympic Legacy Park 140, 141, 142, 143, 145, 284, 285
UTC South Durham 257, 284, 285
UTC Swindon 76, 103, 119, 284, 286
UTC Warrington 284, 285

Vardy, Peter 9, 29
Verma, Baroness 41
VEX Robotics 60
Visions Learning Trust 152
Volvo 9

Wade, Jim 3, 12, 13, 14, 15, 16, 46, 78, 168, 177, 178, 181, 182, 254, 264
Wales, Prince of 17
Walsall College 46, 61, 63, 64
Walsall Education Welfare Service 67
Ward, Nigel 90, 93, 96
Ware, Jane 17, 18, 19, 20, 21, 45
Warwick Manufacturing Group 173
Warwick UTC 77
Water Aid 199
Waterfront UTC 284, 286
Watford UTC 118, 283, 286
Wellington College 258
Welsh Joint Education Committee 113
West Midlands Ambulance Service NHS Foundation Trust 160, 197
West Midlands Construction UTC 77, 117, 160, 245, 248
Wiborg, Susanne 46
Wigan UTC 73, 119, 129, 152, 160, 161, 237, 284
Wilby, Peter 163
Willetts, David 40
Williamson, Gavin 242
Wilson, Rob 98, 99, 100, 101, 108, 109

WiSET 153
WMG Academy for Young Engineers 119, 257, 284, 285
Wokingham Borough Council 99, 100
Wolf, Alison 50, 51, 55, 153
Wolf Report on Vocational Education 50, 51, 55, 166
Wolverhampton University 61, 63, 64, 65, 160, 197
Women into Science and Engineering (WISE) 149, 150, 153, 154, 155, 157, 159
Women's Engineering Society 153
WorldSkills UK 87, 137, 139

Worsey, Leila 213, 214
Wright, Mike 257
Wright, Richard 135, 137
W S Atkins 45
Wylie, Peter 171, 222, 224, 254

York, Duke of 68, 78, 79, 80, 81, 83, 84, 85, 87, 88, 193, 195
Young Apprenticeship scheme 206
Young People's Learning Agency 55, 75

Zambellas, Admiral Sir George 195
ZF Lemförder 64, 65, 69, 86
Zytek Automotive 5